Studies in Systems, Decision and Control

Volume 178

The series "Studies in Systems, Decision and Control" (SSDC) covers both new developments and advances, as well as the state of the art, in the various areas of broadly perceived systems, decision making and control–quickly, up to date and with a high quality. The intent is to cover the theory, applications, and perspectives on the state of the art and future developments relevant to systems, decision making, control, complex processes and related areas, as embedded in the fields of engineering, computer science, physics, economics, social and life sciences, as well as the paradigms and methodologies behind them. The series contains monographs, textbooks, lecture notes and edited volumes in systems, decision making and control spanning the areas of Cyber-Physical Systems, Autonomous Systems, Sensor Networks, Control Systems, Energy Systems, Automotive Systems, Biological Systems, Vehicular Networking and Connected Vehicles, Aerospace Systems, Automation, Manufacturing, Smart Grids, Nonlinear Systems, Power Systems, Robotics, Social Systems, Economic Systems and other. Of particular value to both the contributors and the readership are the short publication timeframe and the world-wide distribution and exposure which enable both a wide and rapid dissemination of research output.

More information about this series at http://www.springer.com/series/13304

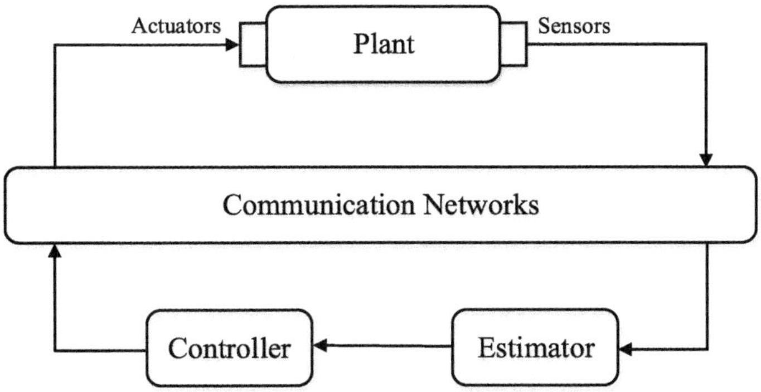

Fig. 1.1 A typical architecture of NCSs

1.1.2 Constrained Communication Networks

Traditional control theory ignores the influence of communication networks or assumes that the communication networks have sufficient large bandwidth so that the communication and control can be treated as two independent processes. Much research effort has been devoted to the fundamental issues such as stability, optimality, resilience, robustness, and decentralization. However, in the emerging applications, such as mobile sensor networks, the signals are sent through wireless channels whose communication bandwidth is quite limited. As a result, the transmitted signals may be distorted, delayed, lost, and sometimes even not be allowed for transmission. All these problems challenge the validity of the traditional control theory in NCSs. Therefore, the NCSs with constrained communication networks have received increasing attention, and more and more researchers have begun to investigate the impact of communication networks on control theory in recent years. According to the available results in the literature, some specific phenomena induced by constrained networks are listed in the following context and their corresponding mathematical models are shown in Table 1.1.

- **Signal quantization**. In NCSs, data is sent through digital communication networks in the form of packets. Usually, the analog signals must be transformed into digital ones by quantizers before transmission. Due to the bandwidth limits, the quantized signals are with a finite word length, which certainly introduces some degree of information loss in the communication process.
- **Communication delay**. Date exchange among the components connected through communication networks suffers from communication delays, which are composed of the sensor-to-estimator delay and the controller-to-actuator delay. The length of time delays is determined by various factors, including the processing speed of devices, the routing algorithm, the network burden, and so on.
- **Packet dropout**. Packet dropout is typically caused by network congestion. The data packets must travel through multiple devices and links to their destination.

Chapter 1
Introduction

1.1 Concepts and Challenges in Networked Control Systems

1.1.1 Networked Control Systems

Networked control systems (NCSs) are spatially distributed systems in which signals between the system components (e.g., sensors, controllers, and actuators) are transmitted via a communication network. Thanks to the rapid development of communication technologies and the increasing demand of remote control, NCSs have gained considerable research interests in the circles of both industry and academia for decades. A typical architecture of NCSs is shown in Fig. 1.1. The sensor devices first convert a physical stimulus of the plant that we are interested in into a readable measurement, and then transmit the measurement to estimators to restore the internal state of the plant. Based on the estimates, the controllers generate the control input and then send it to the actuators to control the plant. That is, the closed-loop control is implemented with the help of communication networks. Using shared communication networks to transfer measurements from sensors to estimators, and control input from controllers to actuators, can greatly ease the complexity of connections, provide more flexibility in architectures, and reduce maintenance costs and troubleshooting. As such, NCSs have found a wide range of applications in engineering practice including mobile sensor networks [1, 2], remote surgery [3, 4], remote diagnostics [5, 6], process control engineering [7, 8], power systems [9, 10], and unmanned aerial vehicles [11, 12].

Q. Liu et al., *Stochastic Control and Filtering over Constrained Communication Networks*, Studies in Systems, Decision and Control 178,
https://doi.org/10.1007/978-3-030-00157-5_1

List of Figures

$\text{diag}\{x_1, x_2\}$	The block diagonal matrix with ith block being x_i and all other entries being zero
$\text{diag}_n\{A_i\}$	The block diagonal matrix $\text{diag}\{A_1, A_2, \cdots, A_n\}$
$\{M_{ij}\}_{n\times n}$	The partitioned matrix with M_{ij} being (i,j)-th block submatrix
$\mathcal{L}_2([0, T]; \mathbb{R}^n)$	The space of square-summable n-dimensional vector functions over [0, T]

Abbreviations

$\odot$	The Hadamard product
$\otimes$	The Kronecker product
$\mathbb{R}^n$	The n-dimensional Euclidean space
$\mathbb{R}^{n\times m}$	The set of all $n \times m$ real matrices
$\mathbb{R}^+$	The set of all positive real numbers
$\mathbb{N}$	The set of natural numbers
$\mathbb{S}^n_+$	The set of $n \times m$ positive-definite matrices
A^T or A'	The transpose of matrix A
$A^\dagger$	The Moore–Penrose pseudoinverse of matrix A
$A > 0$	The matrix A is positive definite
$A \geq 0$	The matrix A is positive semidefinite
$A < 0$	The matrix A is negative definite
$A \leq 0$	The matrix A is negative semidefinite
$\Vert \cdot \Vert$	The Euclidian norm of real vectors or the spectral norm of real matrices
$\Vert x \Vert_M^2$	The quadratic form $x'Mx$
$\Vert A \Vert_{min}$	The smallest singular value of matrix A
$\lambda_{max}(A)$	The eigenvalue of matrix A with the largest modulus
$\lambda_2(A)$	The eigenvalue of matrix A with the second largest modulus
$\mathrm{tr}(A)$	The trace of matrix A
$\mathbb{P}\{\cdot\}$	The occurrence probability of the event "·"
$\mathbb{E}\{x\}$	The expectation of stochastic variable x
$\mathrm{Var}\{x\}$	The variance of stochastic variable x
$\mathbb{E}\{x\|y\}$	The conditional expectation of x given y
I	The identity matrix of compatible dimension
$\mathbf{0}_n$	The $n \times n$ zero matrix
$\mathbf{1}_n$	The $n \times 1$ column vector with all elements equal to 1
$\mathrm{vec}\{x_1, x_2\}$	The column vector $\left[x_1^T x_2^T\right]^T$
$\mathrm{vec}_n\{x_i\}$	The column vector $\mathrm{vec}\{x_1^T, x_2^T, \cdots x_n^T\}$

Contents

Acknowledgements

This book would not have been possible without the help, support, and guidance of many people. The authors would like to express their deep appreciation to those who have been directly involved in various aspects of the research leading to this book.

Special thanks go to Professor Donghua Zhou from Tsinghua University, and Professor Li Qiu and Research Assistant Professor Wei Chen from the Hong Kong University of Science and Technology for their valuable suggestions, constructive comments, and support. The authors also extend our thanks to many colleagues who have offered support and encouragement throughout this research effort. In particular, we would like to acknowledge the contributions from Bo Shen, Guoliang Wei, Hongli Dong, Lifeng Ma, Derui Ding, Jun Hu, Liang Hu, Jinling Liang, Yang Liu, Yurong Liu, Nianyin Zeng, Sunjie Zhang, and Lei Zou. Last but not the least, the authors are especially grateful to their families for their encouragement and never-ending support when it was most required.

The writing of this book was supported in part by the National Natural Science Foundation of China under Grants 61490701, 61525305, 61473163, 61522309, 61873148, 61803282, and 61733009, the Royal Society of the UK, the Research Fund for the Taishan Scholar Project of Shandong Province of China, and the Alexander von Humboldt Foundation of Germany. The work of Xiao He was also partially supported by the R&D Project of Intelligent Ship 1.0 from China's Ministry of Industry and Information Technology under Grant [2016]544, the Special Fund of Suzhou-Tsinghua Innovation Leading Action under Grant 2016SZ0202, and the National Key Research and Development Program of China under Grant 2017YFA0700300. The support of these organizations is gratefully acknowledged. The third author also expresses his acknowledgement for the support from Beijing Association of Automation.

Shanghai, China — Qinyuan Liu
Uxbridge, UK — Zidong Wang
Beijing, China — Xiao He
January 2018

techniques including recursive Riccati equations, matrix decomposition, optimal estimation theory, and mathematical optimization methods are employed to develop the controller/filters with specific performance. In addition, this book provides valuable reference materials for researchers who wish to explore the area of control and filtering under constrained communication networks.

The compendious frame and description of the book are given as follows. Chapter 1 introduces the recent advances on stochastic control and filtering problems for networked systems and the outline of the book. Chapter 2 is concerned with the state feedback stabilization for NCSs subject to signal-to-noise constrained fading channels. Chapter 3 studies the H_∞ consensus control of multi-agent systems, where an event-based mechanism is introduced to reduce the consumption of network resources. Chapter 4 considers the event-based filter design problem for a class of discrete-time stochastic systems. The resilience issue is taken into account in order to accommodate the possible gain variations in the course of filter implementation. Chapter 5 deals with the event-based distributed filtering over wireless sensor networks with bandwidth and energy constraints, and establishes the sufficient conditions for the convergence of the filtering error systems. Chapter 6 extends the results in Chap. 5 to the wireless sensor networks with Markovian switching topologies. Chapter 7 addresses the minimum-variance recursive distributed filter design problems under event-based communication strategies, based on which Chap. 8 further considers the resilience and stability issues. In Chap. 9, a consensus-based distributed filter is developed for a class of discrete time-varying systems subject to stochastic nonlinearities. Chapter 10 discusses the convergence issues for consensus-based distributed filter subject to random link failures. Chapter 11 is concerned with the moving-horizon state estimation problems for a class of discrete-time complex networks under binary encoding schemes. Chapter 12 gives the conclusion and some possible future research topics. Simulations presented in this book are implemented using The MathWorks, Inc. MATLAB software package.

This book is a research monograph whose intended audience is graduate and postgraduate students as well as researchers. The background required of the reader is knowledge of basic stochastic process, basic control system theory, basic Lyapunov stability theory, and basic optimal estimation theory.

Shanghai, China — Qinyuan Liu
Uxbridge, UK — Zidong Wang
Beijing, China — Xiao He
January 2018

Preface

Network control systems (NCSs) are spatially distributed systems whose components, such as actuators, sensors, and controllers, are connected via a shared communication network. Due to the advantages of flexible architectures, reduced installation and maintenance cost, and high reliability, NCSs have attracted increasing attention from researchers and have been widely applied in a broad range of fields such as unmanned aerial vehicles, intelligent building, remote surgery, and automated highway systems. However, it should be stressed that the communication networks in practice are usually resource-constrained, and therefore, some network-induced phenomena would inevitably emerge during the signal transmission, which include, but are not limited to, communication rate constraints, signal-to-noise constraints, channel fading, communication delays, packet dropout, signal quantization, and so on. These phenomena have a great impact on the signals transmitted through communication networks making executors (or estimators) unaware of the exact signals from controller (or sensor). In this case, the performance of NCSs can no longer be guaranteed when utilizing the traditional estimation/filtering techniques. Consequently, it is of practical significance to establish new techniques for control and filtering of the networked system under constrained communication networks. Generally speaking, the research on this topic is concerned with the interplay between three realms: control theory, communication theory, and information theory. The main challenge lies in the signal distortions and the information constraints in the communication loop due to imperfect network conditions.

The objective of this book is to present the up-to-date research developments and novel methodologies on stochastic control and filtering for networked systems under constrained communication networks. The content of this book can be divided into two parts, where the first part (Chaps. 2 and 3) presents control design methodologies and the second part (Chaps. 4–11) shows the filter design methodologies. This work provides a framework of optimal controller/filter design, resilience operation, stability analysis, and performance analysis for the considered systems subject to various kinds of communication constraints such as signal-to-noise constraints, bandwidth constraints, packet drops, etc. Several

Qinyuan Liu
Department of Computer Science and Technology
Tongji University
Shanghai, China

Xiao He
Department of Automation
Tsinghua University
Beijing, China

Zidong Wang
Department of Computer Science
Brunel University London
Uxbridge, UK

ISSN 2198-4182 ISSN 2198-4190 (electronic)
Studies in Systems, Decision and Control
ISBN 978-3-030-00156-8 ISBN 978-3-030-00157-5 (eBook)
https://doi.org/10.1007/978-3-030-00157-5

Library of Congress Control Number: 2018953301

This Springer imprint is published by the registered company Springer Nature Switzerland AG
The registered company address is: Gewerbestrasse 11, 6330 Cham, Switzerland

Qinyuan Liu · Zidong Wang
Xiao He

Stochastic Control and Filtering over Constrained Communication Networks

Springer

Table 1.1 Mathematical models of some phenomena induced by constrained networks

Types	Mathematical models
Signal quantization	There are various types of quantizers. Here, we take a uniform quantizer for an example. The set of quantization levels for the quantizer is described by $$\mathcal{U} = \{\tau_t \| \tau_t \triangleq t\Delta,\ t = 0, \pm 1, \pm 2, \ldots\},\ \Delta > 0. \quad (1.1)$$ The quantization function $Q(.)$ maps the transmitted signal $y \in \mathbb{R}$ into the set $\mathcal{U}$. When $\tau_t \le y(k) \le \tau_{t+1}$, the signal $Q(y) = \tau_t$
Communication delay	Denote the transmitted signal and received signal at instant k by $y(k)$ and $\hat{y}(k)$, respectively, and then the communication delay can be represented by $\hat{y}(k) = y(k - \tau_k)$ with τ_k being the time delay
Packet dropout	A Bernoulli random variable γ_k can be utilized to characterize the packet dropout at instant k. If $\gamma_k = 0$, the packet dropout occurs, and if $\gamma_k = 1$, the packet is sent successfully
SNR constraints	The SNR is considered when the distribution of all signals converges to a stationary distribution. In this case, the power of the transmitted signal y and the channel noise are defined by $\|y\|_{Pow}$ and Φ. Under the SNR constraints, it has $$SNR = \frac{\|y\|_{Pow}}{\Phi} \le \frac{P}{\Phi} \quad (1.2)$$ where P is the admissible transmission power level

When the number of data packets in the transmission sequence overalls the maximum load of the network, network congestion happens, which may result in packet dropout.

- **Signal-to-noise ratio constraints**. Signal-to-noise ratio (SNR), which is defined as the ratio of signal power to noise power, is a measure of the signal strength relative to the background noise in the communication process. For a specific communication network, the data transfer is subject to certain SNR constraints due to the transmission power limits.
- **Bandwidth and energy constraints**. Limited by the capacity of network devices, the bandwidth of communication networks and the energy of transistors are usually restricted.

Generally speaking, the research on NCSs is concerned with the interplay between three realms: control theory, communication theory, and information theory. The main challenge lies in the signal distortions and the information constraints in the communication loop due to imperfect network conditions.

1.2 Analysis and Synthesis of Networked Control Systems

In this chapter, the research on analysis and synthesis of NCSs under four different network-induced effects, i.e., signal quantization, communication delay, packet dropout, and SNR constraints, will be reviewed.

1.2.1 Signal Quantization

As signal quantization is impossible to be performed with infinite precision, it inevitably leads to quantization errors acting on the transmitted signals. Therefore, it is of practical importance to carry out analysis on the quantizers and investigate whether the quantized control/filtering systems can achieve the desired performance. The studies of the signal quantization date back to 1956, when Kalman pointed out that, under quantization schemes, the controlled output might present complicated behaviors such as limit circles and chaos [13]. Since then, the quantized networked systems have gained more and more research attention, and many results have been presented in the literature. Generally speaking, quantizers can be classified into two categories, that is, static quantizer and dynamic quantizer. A static quantizer, in fact, is a memoryless nonlinear function mapping the signals into a quantization set. According to the types of nonlinear functions, the static quantizer can be further classified into several categories including the uniform quantizer, the probabilistic uniform quantizer, the logarithmic quantizer, etc.

As can be seen in (1.1), the quantization levels of a uniform quantizer are equal, and, therefore, the corresponding quantization error affecting the received data can be described as an additive noise with bounded norm. In [14], the tracking problems of an uncertain LTI system with uniformly quantized control input have been studied. A study of visual tracking control of a wheeled mobile robot subject to velocity quantization uncertainties has been presented in [15], where a robust control law that overcomes the unmodeled quantization effect is established. The authors in [16] have considered the cooperative control under uniformly quantized information for multi-agent systems, and have shown that the dynamics of agents will enter a ball when there is a tree in the communication topology.

It should be stressed that, with the uniform quantizer, the asymptotic behaviors of the NCSs cannot be guaranteed [17, 18], so a probabilistic uniform quantizer has been introduced to deal with such an issue. The probabilistic uniform quantizer has the same quantization set as the uniform quantizer, while the major difference lies in that the quantizer function of the former is a stochastic one. To be specific, in a probabilistic uniform quantizer, the quantizer output can be defined as follows:

$$\begin{cases} \mathbb{P}\{\mathcal{Q}(y) = \tau_t\} = 1 - r \\ \mathbb{P}\{\mathcal{Q}(y) = \tau_{t+1}\} = r, \end{cases} \tag{1.3}$$

where $r = \frac{y-\tau_t}{\delta} \in [0, 1]$. In this case, the quantization error, defined by $e = y - Q(y)$, is an additive random variable with zero mean and bounded variance. In [19], a consensus protocol has been proposed in which the agents employ probabilistically quantized information to communicate with the neighboring nodes; it has been proved that consensus can be achieved with the expected consensus value equal to the average of the initial values of agents. Furthermore, a mean-squared error convergence analysis of the consensus error has been proved as time goes to infinity in [20].

The (probabilistic) uniform quantizer uniformly divides the whole segment into equal quantization levels. As compared to the uniform quantizer, the relation between two quantization levels in the logarithmic quantizer is logarithmic. It has been shown in [21] that the logarithmic quantizer is the coarsest quantizer that can quadratically stabilize a single-input linear discrete-time invariant system, and hence the logarithmic quantizer is more preferable in NCSs since it greatly reduces the transmitted bits. In a typical logarithmic quantizer, the set of quantization levels is described by

$$\mathcal{U} = \{\pm\tau_t | \tau_t \triangleq \rho^t \tau_0,\ t = 0, \pm 1, \pm 2, \ldots\} \bigcup \{\pm 0\} \bigcup \{\pm\tau_0\}, \tag{1.4}$$

where $0 < \rho < 1$, $\tau_0 > 0$. The quantized function $Q(\cdot)$ is usually symmetric, i.e., $Q(y) = -Q(-y)$, with the following form:

$$Q(y) = \begin{cases} \tau_t, & \frac{1}{1+\delta}\tau_t \leq y \leq \frac{1}{1-\delta}\tau_t \\ 0, & y = 0 \\ -Q(-y), & y < 0 \end{cases} \tag{1.5}$$

where $\delta = (1-\rho)/(1+\rho)$. According to the results in [22], the logarithmic quantization effects can be described by sector bound uncertainties, that is, $Q(y) = (1+\Delta)y$ for uncertainties Δ satisfying $\|\Delta\| \leq \delta$. The robust analysis methods can be applied to study the stability of quantized control/filtering problems, see, e.g., [23–25] and the references therein. For instance, in [23], Fu and Souza have considered the state estimation problem for networked systems with quantized measurements. The results have revealed the relationship between quantization density and the asymptotic convergence of estimation error variance. The quantized feedback control problem has been considered in [25], where the authors have utilized a quantization dependent Lyapunov function to establish a sufficient condition for stabilization in terms of linear matrix inequalities (LMIs).

The dynamic quantizer can automatically adjust the quantization policy based on the current and the historical data, and thus is more complicated but effective than the static one. A distinct advantage of the dynamic quantizer is that it can stabilize the networked system by using a finite number of quantization levels [26, 27]. In [27, 28], a novel quantized control approach has been proposed, which allows the change of quantizer sensitivity when the system evolves. In this setup, the dynamic quantizer scheme can be divided into two stages: zoom-out stage and zoom-in stage. At the zoom- out stage, the quantization level is increased until the quantizer input can be adequately measured, and subsequently, at zoom-in stage, the quantization level is decreased to drive the state of the plant to zero. Another type of dynamic quantizers is based on the dynamic scaling of quantizer input. That is, the input signal is pre-scaled by the quantizer to make its range be suitable for quantization [26].

1.2.2 Communication Delay

Since the bandwidth of networks and the capacity of transmission devices are both limited, time delays are inevitable in the communication process. The length of communication delays is usually determined by various factors, such as the processing speed of devices, the routing algorithm, and the network burden. In many situations, communication delay is a source of instability [29], and hence it is of theoretical and practical importance to take the communication delays into consideration in the analysis and synthesis of the NCSs. According to the available results in the literature, the communication delays can be described by two models: constant delay and time-varying delay.

The constant delay is the simplest model of the communication delay, but it indeed can be utilized to describe a class of practical applications, wherein the buffers in the receiver is longer than the worst-case delay time [30]. A drawback of such a treatment is its conservatism as the signal delay becomes longer than its actual transfer delay. The analysis and synthesis issues of the NCSs with constant delay have been studied in [31], where the authors give the stability region of the closed-loop systems. For consensus issues of multi-agent systems with constant communication delay, a sufficient condition for consensus has been presented in the form of linear matrix inequalities in [32]. Further studies in [33] have pointed out that for first-order multi-agent systems, the constant communication delay does not affect the consensusability.

Because of stochastic and time-varying properties of network environment, the constant delay is insufficient to characterize the practical communication delay in reality. Recently, significant research effort has been devoted to the analysis and synthesis of networked systems with time-varying delays, which can be modeled in various ways according to the prior knowledge of the communication delay. A simplest way to describe time-varying delays is treating them as an uncertainty with given lower and upper bounds, e.g., the communication delay τ_k satisfies

$$\underline{\tau} \leq \tau_k \leq \bar{\tau}, \quad \text{for } \underline{\tau}, \bar{\tau} \geq 0.$$

In [34], Yue et al. have established the sufficient condition for the asymptotic stability of NCSs with bounded delay by using an linear matrix inequality approach. In [35], the remote state estimation for networked systems with bounded delays have been investigated. The exponential synchronization for complex networks with bounded communication delays has been discussed in [36].

Another way to describe time-varying delays is treating them as random variables. For example, in [37], the delays have been assumed to be a random variable with known statistical properties, and the "probabilistic delay averaging" approach has been utilized to deal with the optimal control issues. In [38], Nilsson et al. have considered that the time delays are independent and their probability distributions are known a priori. An optimal stochastic control scheme has been proposed when the timestamps of the delays are available. In [39], a binary delay has been employed to describe the random communication delay by using a Bernoulli- distributed white sequence, and an H_∞ control method has been established for both the exponential stability and the H_∞ performance of the networked system.

It is worth noticing that, in the aforementioned random delays, the communication delays are assumed to be statistically independent from transfer to transfer. This assumption, however, is quite restrictive in the practical communication network, since the network environment usually changes with a slower time constant than the communication period of the networked systems. That is to say, the delays of two successive transmissions are highly correlated to each other. A way to reflect such correlations between the transfer delays is using the Markov chain. To be specific, the communication delay τ_k is modeled as a homogeneous Markov chain, which takes values in a finite set $M = \{0, 1, \ldots, n\}$ with the transition probability matrix being $\Pi = [\pi_{ij}]_{n\times n}$, i.e.,

$$\pi_{ij} = \mathbb{P}(\tau_{k+1} = j|\tau_k = i), \quad \text{and} \quad \sum_{j=1}^{n} \pi_{ij} = 1, \quad \text{for } i, j \in M.$$

In [40], the stabilization problem for NCSs with Markov delays in both sensor-to-controller and controller-to-actuator loops has been investigated. By transforming the closed-loop systems into Makrov jump linear systems with two modes, the necessary and sufficient conditions on the existence of stabilizing controllers have been established. Further studies in [41] have extended the results in [40] to a more general output feedback stabilization problem. In [6], the fault detection problem has been addressed for a class of networked systems with Markov communication delays, and a robust fault filter with desired H_∞ performance has been designed.

1.2.3 Packet Dropout

Another significant challenge in NCSs is that the data packets may be dropped or lost while transmitted over the communication network. The packet dropout can be caused by a variety of reasons, for instance, an excessively long transmission delay, packet reordering, the disconnection of networks, transmission errors, and so on. The research interest on this topic usually focuses on the modeling of packet dropouts, the optimality of the controller/filter, and the stability and stabilization of networked systems. An arguably popular approach to describe packet dropouts is employing a binary random sequence (i.e., $\{\gamma_k\}_{k\geq 0}$, $\gamma_k \in \{0, 1\}$) that is specified by certain distributions such as a Bernoulli distribution [42, 43] and a Markov process [44]. The related results in the literature can be classified in the light of whether the arrival of the packets is available or unavailable to the receiver.

For the case where packet arrivals are unknown, the statistical properties of the sequence $\{\gamma_k\}_{k\geq 0}$ is utilized in the design of controller/filter with desired performance. For example, in [42], Nahi has addressed the state estimation problems subject to missing measurements. The measurements are affected by a Bernoulli variable with the probability of occurrence being available to the estimator. Based on the first- and second-order moments of the random variable, an optimal recursive estimation algorithm in the minimum mean-square error sense has been developed. In [45], the output feedback control problem for networked systems with random packet dropouts has been addressed. The authors have constructed a dynamic observer-based control scheme by using the expectation of the Bernoulli variable that describes the packet dropouts, and established sufficient conditions for the existence of the robust H_∞ controller. Moreover, in [44], a Markov process has been utilized to model the packet dropouts and the output power of NCSs has been shown as a function of the transition matrix of the Markov process. One distinct advantage of treating the packet arrivals as unknown variables lies in that the design of optimal controller/filter can be performed in an off-line fashion. That is, the controller/filter parameters can be established by using the statistical properties, regardless of the situation of real-time packet dropouts. It should be pointed out that in this case, the obtained controllers/filters are robust over the packet dropouts, but sometimes the stability of controlled systems or estimation error systems is highly related to the stability of the open-loop systems [43].

Another framework to deal with the packet dropouts is regarding them as a sequence with known values in a single sampling path. For example, in [46], Sinopoli et al. have considered a linear discrete-time system with the arrivals of measurements modeled as i.i.d. Bernoulli variables and proposed an optimal filter with packet dropouts by inserting the sequence $\{\gamma_k\}_{k\geq 0}$ into the traditional Kalman filter. To be specific, when the packet arrives, the Kalman filter is adopted to obtain the optimal estimate, while when the packet is dropped, the estimator utilizes the model of the plant to generate a one-step prediction. As such, different from the Kalman filter, the dynamics of error covariance is now a random Riccati equation depending on $\{\gamma_k\}_{k\geq 0}$, which can only be calculated online. It has been shown that there exists a critical value of the dropout probability below which the mean of the error covariance is unbounded. Subsequent work in [47] has established a relationship between such a critical value and the spectral radius of the transition matrix for a class of linear discrete-time systems. The weak convergence of the random error covariance matrices to a unique invariant distribution has been investigated in [48]. Furthermore, in [49], Schenato et al. have addressed the LQG problems for the NCSs where the sensor signals and the control signals may be lost. Both the TCP-like and UDP-like communication protocols have been taken into consideration. In TCP-like protocols, whether the transmissions of control signals are successful will be provided to controller, while in UDP-like protocols, there is no such feedback. It has been shown that under TCP-like protocols the separation principle holds, which means that controller design and estimator design are decoupled, and the expectation of estimator error covariance is bounded if and only if dropout probability exceeds a certain threshold. In [50], the stability of Kalman filtering over Gilbert-Elliott channels with the random packet dropout described by a two-state Markov chain has been addressed. The results have proved that there exists a critical curve in the failure-recovery rate plane, below which the proposed filter is mean-square stable.

1.2.4 SNR Constraints

The quality of transmission highly relies on the ratio between the energy of the additive noise and that of the channel input. Due to the constraints of equipment and technologies, the power of the transmitted signal is usually constrained. Therefore, it is of practical importance to investigate the influence of SNR constraints on the performance of NCSs. Up to now, considerable research attention has been devoted to this topic in recent years. For example, in [51, 52], the feedback stabilization problem has been studied for single-input and single-output NCSs over additive noise channels with power constraints. Necessary and sufficient conditions for the networked stabilization have been established, which clearly reveals that the minimum SNR required for stabilization shall be larger than a function of unstable plant poles under a static state feedback scheme, while a function of the non-minimum phase zeros of the plant under a dynamic output feedback scheme. For multi-input systems, the minimum SNR requirement issue for the networked stabilization problem is getting more complicated as compared to the single-input case, and an explicit expression for the SNR limitation on multi-input networked stabilization is difficult to obtain. Some promising progresses have been made in [53–57]. For example, the stabilization of discrete-time multi-input multi-output (MIMO) systems over parallel communication channels subject to independent SNR constraints has been addressed in [56, 57]. Necessary and sufficient conditions for stabilizability have been derived involving both the values and directions of the unstable poles. Recently, a line of research [24, 58, 59] utilized the idea of channel resource allocation to find the infimum SNR required for stabilizability. By virtue of the additional design freedom due to channel resource allocation, it has been shown that the minimum total SNR required for multi-input networked stabilization can be characterized by the topological entropy of an open-loop plant.

1.3 Event-Based Control and Filtering Problems

As is well known, an efficient method to reduce the bandwidth and energy consumption in the communication process is utilizing event-based strategies. Under the event-based strategies, a scheduler is employed in the transmitter to determine whether the current signal shall be transmitted or not. In this way, the time intervals between two successive transmissions will be increased to a great extent. In this chapter, the results on the event-based control and filtering in the literature will be reviewed.

1.3.1 Event-Based Control for Networked Control Systems

To deal with event-based control strategies for linear NCSs, several different analysis approaches have been proposed in [60, 61] as follows: (1) an impulsive system approach; (2) a discrete-time piecewise linear (PWL) system approach; and (3) a discrete-time perturbed linear (PL) system approach. The impulsive system approach exactly reveals the inter-sample behaviors of event-based control, and the PWL system approach models the controlled system as piecewise functions. The PL system approach regards the control errors introduced by events as a disturbance of the nominal system, and thus it naturally resorts to the perturbation method for further analysis [62]. Most of the existing results have been based on the PL approach, see e.g., [63–65] and so on. However, it is still noteworthy that, compared with the first two approaches, this PL system approach usually establishes a more conservative stability condition. In addition, in the work [66, 67], for the convenience of analysis and synthesis, a delay system method has been proposed to investigate the networked control systems with transmission delays under the event-based scheme. Unlike the aforementioned works [63–65, 68], the event generator monitors the triggering condition in a discrete-time manner instead of a continuous one. Since the lowest execution intervals are determined by discrete-time supervision instants, there is neither accumulation point nor Zeno-sampling phenomenon during the running process of the system.

In [69, 70], the authors have presented preliminary results of control synthesis for systems with event-based schemes, where the controllers have been designed to ensure the asymptotic/exponential stability of the closed-loop systems under the assumption of input-to-state stability with respect to measurement errors. In [71], a schedule called fixed threshold strategy(FTS) has been proposed to guarantee the bounded-input bounded-output (BIBO) stability of the controlled system, where the switching boundary is determined a priori as a fixed positive constant. When the boundary is set to zero, the event-based approach will reduce to the time-driven one. Furthermore, a relative threshold strategy (RTS) has been investigated to adjust the diameter of the boundary in proportion to the norm of the system state. It is clear that the controlled system is asymptotically stable and the plant state will be driven to a decreasing area. Many established results referring to event-triggering schemes are in the framework of continuous-time systems, while some results have been extended to control discrete-time systems, see [72]. In fact, such a discrete-time controlled system does have its inherent benefit, for instance, the lower bound of execution intervals would be no shorter than the sampling period. However, for those continuous-time systems, the desired lower bound on execution interval (if it exists) is rather difficult to design under the complicated impacts of the system parameter settings.

It is worth mentioning that, there usually exist model parameter uncertainties in many practical applications. Therefore, in [65, 73, 74], some researchers have proposed model-based periodic event-based control strategies for discrete-time linear plants subject to disturbances and model uncertainties. The model-based technologies, which construct model-based predictors running both at the sensor and the controller system, have been well recognized as an effective method in reducing the impact of model uncertainties in absence of continuous feedback signals. Compared with the traditional open-loop event-based control, in this case, continuous closed-loop signals for real-time control can be generated using predictors with aperiodic updates. Furthermore, in [61] the authors have investigated the model-based periodic event-triggered mechanism under two general analysis frameworks of PL systems and PWL systems.

Over the past 10 years or so, event-based control of distributed NCSs has gained increasing attention from researchers. It is obvious that, in the centralized triggering schemes, the global information is demanded for event generators to determine synchronous execution instants, which may lead to the overburden of communication networks as well as a waste of energy. Some alternative event-based distributed strategies have been proposed to alleviate the occupation of network, see, for example, the works in [60, 69, 75–77]. Nevertheless, we can see that the decentralized triggering conditions are more conservative than the centralized ones, namely, the average execution intervals is usually shorter in decentralized case. Consequently, a zero-sum set of online adaption parameters scheme has been proposed to reduce this conservation [76].

1.3.2 Event-Based Consensus Control for Multi-agent Systems

Very recently, the cooperation of multi-agents has been an increasingly popular topic for its potential prospects in variety of realms such as military, industry, etc. Up to now, numerous results concerned with the formulation, rendezvous, leader-following problems in multi-agent systems have been reported. Essentially, the multi-agent systems constitute a subset belonging to the distributed systems, where each intelligent agent utilizes the available information from not only itself but also its neighboring agents for the purpose of collectively performing certain tasks or actions [78]. A paramount problem lies in is how to appropriately design the distributed algorithms for both communication protocol and control scheme. In general, these distributed agents are equipped with small but highly integrated microprocessors. These mini processors guarantee the regular communication between neighbor agents and calculate the control signals for the actuators. In terms of traditional distributed strategies, signal transmissions and updates should be executed at every sampling instant. It increases the risk of some undesired phenomena, including communication congestion as well as frequent modulation of actuators, which might have a destructive impact on the network environments and damage the lifespan of actuators, respectively. Motivated by the discussion above, an alternative aperiodic approach named event-based consensus protocol has been introduced into multi-agent systems with limited resources. In this framework, plenty of redundant executions would be abandoned because the actuators are modulated only when some specific event happens. Hence, it is natural that event-based approaches are more favorable than time-driven ones for actual multi-agent systems.

Generally, the consensus protocol for multi-agent systems calls for the information exchanging between neighboring agents in a periodic mechanism [78]. An important aspect in the implementation of event-based strategies lies in the communication and control schemes. There are two event-based consensus protocols in the existing literature: (1) centralized event-based consensus control; (2) decentralized event-based consensus control. In terms of centralized event-based approaches, there only has a single event generator determining the global triggering instants for each agent to broadcast the local information and update control input at the same time. On the other hand, decentralized approaches require every smart agent to equip event generators in order to transmit their local information to neighborhoods in asynchronous instants rather than the synchronous ones. Besides, after receiving the broadcast, the agents only need to update the corresponding control input from the message sources. It is obvious that the decentralized event-based approaches own the inherent advantages since they do not demand a triggering center to guide the behaviors of the multi-agent systems.

In [79], an event-based communication strategy has been proposed to ensure the target configuration of vehicles, whose motion can be described as a single integrator with external disturbances. Concerning with centralized event-based approaches with the global information, some conclusions have been reached, see [80–82]. In [80], the formulation control of multi-agent systems has been addressed. In [81], linear matrix inequalities have been utilized to investigate the average consensus problem in discrete-time systems with the predefined event generator condition. For second-order multi-agent systems, an event-based mechanism has been presented to guarantee the stability of systems under limited resources in [82]. However, despite the fact that centralized triggering approaches are relatively simple to implement, every agent has to know the global information in order to decide the next triggering instant. Consequently, to address the challenges of reasonable bandwidth allocation, a decentralized event-based control mechanism has been introduced in [83–86]. To be specific, the authors have considered the second-order leader-following multi-agent systems with partial observations in [83]. The velocity of the active leader remains unknown but could be estimated for the following agents. Based on the framework of [83, 84], the event-based control problem has been addressed with partial observations and communication delays, where some partially uncertain acceleration exists in the dynamic system of the active leader. Insights on event-based control for multi-agent systems in a fixed undirected network topology has been proposed in [85], where both centralized and decentralized manners are applied to a first-order agreement problem. Additionally, a self-triggered strategy has been explored to calculate the next sampling instant in absence of the system state. In [86], the authors have discussed the event-based average consensus problem under the assumption of directed and weighted communication topology.

However, it is worth mentioning that, for many established decentralized event-based approaches, the event generator located in each agent utilizes real-time information from its neighboring agents, such that each agent has to broadcast the states at every sampling period to keep all its neighbors informed about its local information. This framework may result in the additional energy consumption of agents and go against the original intentions of event-based approaches. Some appropriate triggering conditions have been established to avoid this unreasonable requirement. Specifically, in [87], the rendezvous problem has been solved by a distributed event-based method, and then an iterative algorithm has been proposed to deal with the continuous measurements of the neighbor states. A novel triggering strategy independent of the real-time state of neighbors has been further investigated in [88] for both single-integrator and double-integrator agents, and, therefore, the continuous monitoring is no longer needed. Nevertheless, a drawback of this strategy is that the global network topology should be available to each agent in order to appropriately design the distributed event generator strategies. Moreover, the consensus problem of discrete-time heterogeneous multi-agent systems of single-integrator and double-integrator agents has been explored under distributed event-based control in [89]. Based on Kronecker product and Lyapunov functional methods, the feedback gain matrices have been designed to guarantee the agreement. Furthermore, in [90], the event-based control of discrete-time heterogeneous multi-agent systems with Markov communication delays has been addressed.

1.3.3 Event-Based Remote State Estimation

Filtering or estimation problems have received lots of attention over the past decades. They have been widely brought into practical applications to restore the state-vector of the plant from observations with external disturbances. In traditional filtering algorithms (e.g., Kalman filter [91]), the synchronous innovations are required in every iteration cycle, so that the fixed sampling period should be designed under the worst conditions. Moreover, with the development of network technologies, nowadays, many industrial manufacturers have introduced the networked environment infrastructures due to their capability of reducing the implementation difficulties and increasing the equipment reliability. System parameters (e.g., temperature, pressure) are usually transmitted to data centers via communication channels to solve the filtering problems in networked systems. However, the waste of bandwidth arises from the transmission of unnecessary information in the time-driven strategy. It is well known that the event-based estimation is a highly effective alternative scheme aiming to enlarge the execution intervals, which has both engineering significance and practical importance. As a result, it has recently become a hot research topic and has attracted an increasing interest in constrained resources systems.

In [92], the triggering mechanism is based on a send-on-delta (SOD) regulation, namely, only when the measurement values change more than a predetermined threshold Δ, the sensors transmit their sampling data to data centers for processing. Besides, the advantages of the SOD concept over a time-driven one has been fully discussed. After that, in [93], an event-based approach has been investigated to estimate the system states with intermittent measurements by a modified Kalman Filter. Furthermore, in [94, 95], the authors have considered the hybrid update strategy to reduce the filtering error of the remote filters without real-time innovations. When there are no new measurements, it can recognize that the difference between the present and the latest measurement values is limited within a bounded subset of the measurement space. Furthermore, the posterior probability of estimation can be updated with such bounded subset. One major issue of the SOD scheme pointed out in [96] is that this fixed threshold strategy cannot detect the steady-state error or small state oscillation, which undesirably reduces the estimation performance. Additionally, since these methods are sensitive to the process noise, which will frequently cause unnecessary triggering, a send-on-area (SOA) triggering approach has been proposed in [97] to overcome these disadvantages by utilizing the integral of such difference instead.

Based on the fundamental ideas above, some papers use standard probability density function to set up a general mathematical description of this event-based hybrid update scheme, see [98–100]. Not confined to the traditional SOD scheme, a more general bounded Borel set in time-measurement-space is under consideration to denote the triggering area [98, 99]. Moreover, a summation of Gaussians has been employed to approximate its probability distribution function so that one can minimize the computational complexity. In [101], a remote-estimator has been put forward to minimize the mean-square error by a sequence of special measurements satisfying triggering rules. Other works that have presented the minimum mean-squared error (MMSE) estimator with an event-based scheme can be found in [98, 100]. Similarly, an approximation technique from nonlinear filtering is exploited to derive an approximate MMSE estimator for computational convenience. Concerning with nonlinear systems, the event-based particle filtering and vector quantization methods have been proposed to estimate the plant's state in [102].

Recently, some results about event-based estimation have been reported under the framework of event-based methods. For networked systems, H_∞ filtering problems have been addressed to estimate the state of the plant in [103], where the new measurements are triggered in an aperiodic manner and transmitted over a bandwidth-limited network with communication delays. The LMIs have been employed to appropriately design the filter parameters that guarantee both the exponential stability and the noise energy attenuation at a certain level. The results have been extended to solve state estimation in delayed recurrent neural networks in [104]. It is well worth mentioning that the triggering mechanism of these two papers is quite different from existing SOD schemes. Since the threshold figuring, the triggering area is tuned according to an adaptive value, one can see that this threshold will decrease along with the decline of estimation errors. Obviously, such an adaptive threshold will ultimately eliminate the negative influences of steady-state error and small state oscillations.

Moreover, the event-based scheme has been further utilized in fault diagnosis in the literature [105], where the authors presented a modified fault isolation filter (FIF) for discrete-time networked control systems with multiple faults. Unlike the traditional time-driven FIFs, the measurements are available for the modified filter in this chapter only when the prescribed events are triggered.

1.4 Outline of This Book

This book is divided into 12 chapters, where Chaps. 2 and 3 mainly focus on the networked control problems for linear systems, and Chaps. 4–11 are concerned with the filtering problems. To be specific, the outline of this book is given as follows:

- This chapter presents some concepts and challenges in NCSs; reviews research progresses of control and filtering problems under the constrained communication networks; and lists the outline of the book.
- Chapter 2 investigates the stabilization problems for NCSs over fading channels. A channel capacity notion is introduced for the fading channels to represent their capabilities of information transmission. A necessary and sufficient condition for closed-loop stabilization with stationary signal-to-noise ratio (SNR) constraints is established which builds a bridge between the channel capacities and the topological entropy of the open-loop plant.
- Chapter 3 considers the H_∞ consensus control problem for general discrete time-varying multi-agent systems with constrained communication bandwidth. A novel event-based control protocol is developed for each agent to reduce the communication burden. A constrained recursive Riccati difference equation approach is employed, which explicitly gives the sufficient conditions for H_∞ consensus performance of the multi-agent systems under event-based schemes. Furthermore, the desired controller and estimator parameters can be iteratively computed by resorting to the Moore–Penrose pseudo inverse.
- Chapter 4 discusses the event-based state estimation problem for discrete-time stochastic systems susceptible to possible sensor failures and signal quantization. The resilience issue for the filter design is taken into account in order to accommodate the possible gain variations in the course of filter implementation. By minimizing the upper bound of the estimation error covariance, an event-based resilient filter is designed. Further analysis shows the monotonicity of the minimized upper bound with respect to the value of thresholds, and establishes the sufficient condition for the convergence of the steady-state filter.
- Chapter 5 deals with the distributed filtering problem with bandwidth and energy constraints. To utilize the limited resources efficiently, a novel event-based mechanism is proposed for the sensor node, in which only a part of the selected data is broadcast. By resorting to the graph theory and stochastic analysis methods, the

filter parameters and the event-triggering rules are designed such that the filtering error converges at an exponential rate in the mean square sense. An adaptive algorithm for determining the triggering threshold is developed to keep the average transmission rate around a desired value.

- Chapter 6 extends the results in Chap. 5 to stochastic systems subject to partial measurements and Markovian switching topologies. A distributed filter that is able to accommodate Markovian switching topologies is designed. It is shown in this chapter that the exponential mean-square boundedness of the estimation errors can be guaranteed if the sensor network is distributively detectable and the combined communication topology is strongly connected.
- Chapter 7 considers minimum mean-square error distributed filter design problems under the event-based communication mechanism. The triggering information is utilized to reduce the conservatism in the filter design. An upper bound for the filtering error covariance is obtained in form of Riccati-like difference equations by utilizing an inductive method. Subsequently, such an upper bound is minimized by appropriately designing the filter parameters iteratively, wherein a novel matrix simplification technique is developed to handle the challenges resulting from the sparseness of the sensor network topology and filter structure preserving issues.
- Chapter 8 further investigates minimum mean-square error distributed filter design problems for the systems with stochastic nonlinearities and sensor degradation. A two-step distributed filter algorithm is proposed where the sensor nodes collaboratively estimate the state of the plant by exploiting the information from both the local and neighboring nodes. A resilient operation is guaranteed to suppress random perturbations on the actually implemented filter gains. An upper bound is first derived for the filtering error covariance by utilizing an inductive method and such an upper bound is subsequently minimized via iteratively solving a quadratic optimization problem. A sufficient condition is established to guarantee the mean-square boundedness of the estimation errors.
- Chapter 9, a two-step consensus-based distributed filtering algorithm is proposed, where the consensus process subject to a given directed graph is proposed to accelerate the information fusion among the local estimates of each sensor. By employing the least square method, the suboptimal filter gain matrices that minimize local error covariance are obtained.
- Chapter 10 addressed the distributed state estimation problem over wireless sensor networks, where the communication links are unreliable that are subject to random link failures modeled as a set of independent Bernoulli processes. To estimate the plant state collaboratively, a Kalman-consensus filtering approach is developed where the sensors spread the local information obtained from the Kalman filtering algorithm by performing a consensus of the inverse covariance matrices at each time instant. Sufficient conditions for the stochastic boundedness of the Kalman-consensus filter are established. It is shown that the filtering performance is directly influenced by the network connectivity and the collective observability.
- Chapter 11 deals with moving-horizon state estimation problems for a class of discrete-time complex networks under binary encoding schemes. A binary encoding scheme, which takes advantages of the robustness of the binary data, is

exploited in signal transmission. Under such a scheme, the original signals will be encoded into a bit string; transmitted via memoryless binary symmetric channels with certain crossover probabilities; and eventually restored by a decoder in the receiver. Novel centralized and decentralized moving-horizon estimators in the preserve of the binary encoding schemes are constructed by solving the respective global and local least square optimization problems. Sufficient conditions for the stochastically ultimate boundedness of the estimation errors have been established.
- Chapter 12 draws the conclusions and points out some potential research directions related to the work done in this book.

In addition, the illustrative numerical simulation examples, which are employed to validate the effectiveness of the results, are provided within the individual chapters.

References

1. G. Hoffmann, C. Tomlin, Mobile sensor network control using mutual information methods and particle filters. IEEE Trans. Autom. Control **55**(1), 32–47 (2010)
2. N. Heo, P. Varshney, An intelligent deployment and clustering algorithm for a distributed mobile sensor network, in *IEEE International Conference of Systems, Man and Cybernetics*, vol. 5 (2003), pp. 4576–4581
3. J. Arata, H. Takahashi, P. Pitakwatchara, S. Warisawa et al., A remote surgery experiment between Japan and Thailand over Internet using a low latency CODEC system, in *IEEE International Conference of Robotics and Automation* (2007), pp. 953–959
4. J. Marescaux, F. Rubino, Robot-assisted remote surgery: technological advances, potential complications, and solutions. Surg. Technol. Int. **12**, 23–26 (2004)
5. S. Brax, K. Jonsson, Developing integrated solution offerings for remote diagnostics: a comparative case study of two manufacturers. Int. J. Oper. Prod. Manag. **29**(5), 539–560 (2009)
6. X. He, Z. Wang, D. Zhou, Robust fault detection for networked systems with communication delay and data missing. Automatica **45**(11), 2634–2639 (2009)
7. N. El-Farra, P. Mhaskar, Special issue on "control of networked and complex process systems". Comput. Chem. Eng. **32**(9), 1963–1963 (2008)
8. Y. Sun, N. El-Farra, Resource aware quasi-decentralized control of networked process systems over wireless sensor networks. Chem. Eng. Sci. **69**(1), 93–106 (2012)
9. A. Casavola, G. Franze, Coordination strategies for networked control systems: a power system application, in *10th International Conference on Control Automation Robotics & Vision*, vol. 1–4 (2008), pp. 503–508
10. P. Park, Power controlled fair access protocol for wireless networked control systems. Wirel. Netw. **21**(5), 1499–1516 (2015)
11. K. Kim, D. Shim, S. Sastry, Nonlinear model predictive tracking control for rotorcraft-based unmanned aerial vehicles, in *Proceedings of American Control Conference* (2002), pp. 3576–3581
12. G. Ducard, Fault-tolerant flight control and guidance systems: practical methods for small unmanned aerial vehicles (Springer Science & Business Media, 2009)
13. R. Kalman, Nonlinear aspects of sampled-data control systems, in *Proceedings of the Symposium on Nonlinear Circuit Analysis*, vol. 6 (1956), pp. 273–313
14. H. Sun, N. Hovakimyan, T. Basar, L_1 adaptive controller for quantized systems, in *Proceedings of American Control Conference* (San Francisco, USA, 2011), pp. 582–587
15. C.Y. Tsai, K.T. Song, Visual tracking control of a wheeled mobile robot with system model and velocity quantization robustness. IEEE Trans. Control Syst. Technol. **17**(3), 520–527 (2009)

16. D.V. Dimarogonas, K.H. Johansson, Quantized agreement under time-varying communication topology, in *Proceedings of American Control Conference* (Seattle, USA, 2008), pp. 4376–4381
17. H. Ishii, T. Basar, Remote control of LTI systems over networks with state quantization. Syst. Control Lett. **54**(1), 15–31 (2005)
18. P. Frasca, R. Carli, F. Fagnani, S. Zampieri, Average consensus on networks with quantized communication. Int. J. Robust Nonlinear Control **19**(16), 1787–1816 (2009)
19. T. Aysal, M. Coates, M. Rabbat, Distributed average consensus using probabilistic quantization, in *IEEE/SP 14th Workshop on Statistical Signal Processing* (2007), pp. 640–644
20. R. Carli, F. Fagnani, P. Frasca, S. Zampieri, Gossip consensus algorithms via quantized communication. Automatica **46**(1), 70–80 (2010)
21. N. Elia, S. Mitter, Stabilization of linear systems with limited information. IEEE Trans. Autom. Control **46**(9), 1384–1400 (2001)
22. M. Fu, L. Xie, The sector bound approach to quantized feedback control. IEEE Trans. Autom. Control **50**(11), 1698–1711 (2005)
23. M. Fu, C.E. de Souza, State estimation for linear discrete-time systems using quantized measurements. Automatica **45**(12), 2937–2945 (2009)
24. L. Qiu, G. Gu, W. Chen, Stabilization of networked multi-input systems with channel resource allocation. IEEE Trans. Autom. Control **58**(3), 554–568 (2013)
25. H. Gao, T. Chen, A new approach to quantized feedback control systems. Automatica **44**, 534–542 (2008)
26. M. Fu, L. Xie, Finite-level quantized feedback control for linear systems. IEEE Trans. Autom. Control **54**(5), 1165–1170 (2009)
27. R. Brockett, D. Liberzon, Quantized feedback stabilization of linear systems. IEEE Trans. Autom. Control **45**(7), 1279–1289 (2000)
28. D. Liberzon, Nonlinear stabilization by hybrid quantized feedback, in Lecture Notes in Computer Science (2000), pp. 243–257
29. J. Hale, S. Lunel, *Introduction to Functional Differential Equations* (Springer, New York, 1993)
30. R. Luck, A. Ray, An observer-based compensator for distributed delays. Automatica **26**(5), 903–908 (1990)
31. W. Zhang, M. Branicky, S. Phillips, Event-triggered tracking control for heterogeneous multi-agent systems with Markov communication delays. IEEE Control Syst. Magine **21**(1), 84–99 (2001)
32. J. Qin, H. Gao, W. Zheng, Second-order consensus for multi-agent systems with switching topology and communication delay. Syst. Control Lett. **60**(6), 390–397 (2011)
33. W. Wang, J.-J. Slotine, Contraction analysis of time-delayed communications and group cooperation. IEEE Trans. Autom. Control **51**(4), 712–717 (2006)
34. D. Yue, Q. Han, C. Peng, State feedback controller design for networked control systems. IEEE Trans. Circuits Syst. **51**(11), 640–644 (2004)
35. L. Zou, Z. Wang, H. Gao, X. Liu, State estimation for discrete-time dynamical networks with time-varying delays and stochastic disturbances under the Round-Robin protocol. IEEE Trans. Neural Netw. Learn. Syst. **28**(5), 1139–1151 (2017)
36. Z. Zuo, C. Yang, Y. Wang, A unified framework of exponential synchronization for complex networks with time-varying delays. Phys. Lett. A **374**(19), 1989–1999 (2010)
37. I.V. Kolmanovsky, T.L. Maizenberg, Optimal control of continuous-time linear systems with a time-varying random delay. Syst. Control Lett. **44**, 119–126 (2001)
38. J. Nilsson, B. Bernhardsson, B. Wittenmark, Stochastic analysis and control of real-time systems with random time delays. Automatica **34**(1), 57–64 (1998)
39. F. Yang, Z. Wang, Y. Hung, M. Gani, H_∞ control for networked systems with random communication delays. IEEE Trans. Autom. Control **51**(3), 511–518 (2006)
40. L. Zhang, Y. Shi, T. Chen, B. Huang, A new method for stabilization of networked control systems with random delays. IEEE Trans. Autom. Control **50**(8), 1177–1181 (2005)
41. Y. Shi, B. Yu, Output feedback stabilization of networked control systems with random delays modeled by Markov chains. IEEE Trans. Autom. Control **54**(7), 1668–1674 (2009)

42. N. Nahi, Optimal recursive estimation with uncertain observation. IEEE Trans. Inf. Theory **IT-15**(4), 457–462 (1969)
43. S. Sun, S.L. Xie, W. Xiao, Y. Soh, Optimal linear estimation for systems with multiple packet dropouts. Automatica **44**(5), 1333–1342 (2008)
44. Q. Ling, M.D. Lemmon, Soft real-time scheduling of networked control systems with dropouts governed by a Markov chain, in *Proceedings of American Control Conference* (Denver, 2003), pp. 4845–4850
45. Z. Wang, F. Yang, D. Ho, X. Liu, Robust H_∞ control for networked systems with random packet losses. IEEE Trans. Syst. Man Cybern. Part B **37**(4), 916–924 (2007)
46. B. Sinopoli, L. Schenato, M. Franceschetti, K. Poolla, M. Jordan, S. Sastry, Kalman filtering with intermittent observations. IEEE Trans. Autom. Control **49**(9), 1453–1464 (2004)
47. Y. Mo, B. Sinopoli, A characterization of the critical value for Kalman filtering with intermittent observations, in *Proceedings of IEEE Conference on Decision Control* (Cancun, Mexico, 2008), pp. 2692–2697
48. S. Kar, B. Sinopoli, J. Moura, Kalman filtering with intermittent observations: weak convergence to a stationary distribution. IEEE Trans. Autom. Control **57**(2), 405–420 (2012)
49. L. Schenato, B. Sinopoli, M. Franceschetti, K. Poolla, S.S. Sastry, Foundations of control and estimation over lossy networks. Proc. IEEE **95**(1), 163–187 (2007)
50. J. Wu, G. Shi, B. Anderson, K. Johansson, Kalman filtering over Gilbert-Elliott channels: stability conditions and critical curve. IEEE Trans. Autom. Control (2017) (available online)
51. J.H. Braslavsky, R.H. Middleton, J.S. Freudenberg, Feedback stabilisation over signal-to-noise ratio constrained channels, in *Proceedings of American Control Conference* (Boston, USA, 2004)
52. J.H. Braslavsky, R.H. Middleton, J.S. Freudenberg, Feedback stabilization over signal-to-noise ratio constrained channels. IEEE Trans. Autom. Control **52**(8), 1391–1403 (2007)
53. T. Tanaka, Semidefinite representation of sequential rate-distortion function for stationary Gauss-Markov processes, in *IEEE Conference on Control Applications* (Sydney, Australia, 2015), pp. 1217–1222
54. T. Tanaka, K. Kim, P. Parrilo, S. Mitter, Semidefinite programming approach to gaussian sequential rate-distortion tradeoffs. IEEE Trans. Autom. Control **62**(4), 1896–1910 (2017)
55. F. Vargas, J. Chen, E. Silva, Necessary and sufficient conditions for mean square stabilization over MIMO SNR-constrained channels, in *2012 IEEE 51st IEEE Conference on Decision and Control* (Maui, USA, 2012), pp. 26–31
56. F. Vargas, E. Silva, J. Chen, Stabilization of two-input two-output systems over SNR-constrained channels. Automatica **49**, 3133–3140 (2013)
57. F. Vargas, E. Silva, J. Chen, Explicit conditions for stabilization over noisy channels subject to SNR constraints, in *Proceedings of Asian Control Conference* (Istanbul, Turkey, 2013), pp. 1–6
58. W. Chen, L. Qiu, Stabilization of networked control systems with multirate sampling. Automatica **49**(6), 1528–1537 (2013)
59. N. Xiao, L. Xie, L. Qiu, Feedback stabilization of discrete-time networked systems over fading channels. IEEE Trans. Autom. Control **57**(9), 2176–2189 (2012)
60. W. Heemels, M. Donkers, A. Teel, Periodic event-triggered control for linear systems. IEEE Trans. Autom. Control **58**(4), 847–861 (2012)
61. W. Heemels, M. Donkers, Model-based periodic event-triggered control for linear systems. Automatica **49**(3), 698–711 (2013)
62. H.K. Khalil, *Nonlinear Systems*, 3rd edn. (Prentice Hall, Upper Saddle River, 2002)
63. P. Tabuada, Event-triggered real-time scheduling of stabilizing control tasks. IEEE Trans. Autom. Control **52**(9), 1680–1685 (2007)
64. X. Meng, T. Chen, Event-based stabilization over networks with transmission delays. J. Control Sci. Eng. **2012**(2) (2012)
65. E. Garcia, P. Antsaklis, Model-based event-triggered control for systems with quantization and time-varying network delays. IEEE Trans. Autom. Control **58**(2), 422–434 (2013)

66. S. Hu, D. Yue, Event-triggered control design of linear networked systems with quantizations. ISA Trans. **51**(1), 153–162 (2012)
67. D. Yue, E. Tian, Q. Han, A delay system method for designing event-triggered controllers of networked control systems. IEEE Trans. Autom. Control **58**(2), 475–481 (2013)
68. A. Molin, S. Hirche, On the optimality of certainty equivalence for event-triggered control systems. IEEE Trans. Autom. Control **58**(2), 470–474 (2013)
69. X. Wang, M. Lemmon, Decentralized event-triggered broadcasts over networked control systems, in *Hybrid Systems: Computation and Control* (Springer, Berlin, 2008), pp. 674–677
70. X. Wang, M. Lemmon, On event design in event-triggered feedback systems. Automatica **47**(10), 2319–2322 (2011)
71. E. Garcia, P. Antsaklis, Model-based event-triggered control with time-varying network delays, in *Proceedings of 50th IEEE Conference on Decision and Control and European Control Conference* (Orlando, USA, 2011), pp. 1650–1655
72. A. Eqtami, D.V. Dimarogonas, K.J. Kyriakopoulos, Event-triggered control for discrete-time systems, in *Proceedings of American Control Conference* (Baltimore, USA, 2010), pp. 4719–4724
73. E. Garcia, P. Antsaklis, Parameter estimation in time-triggered and event-triggered model-based control of uncertain systems. Int. J. Control **85**(9), 1327–1342 (2012)
74. S. Hu, X. Yin, Y. Zhang, E. Tian, Event-triggered guaranteed cost control for uncertain discrete-time networked control systems with time-varying transmission delays. IET Control Theory Appl. **6**(18), 2793–2804 (2012)
75. X. Wang, Y. Sun, N. Hovakimyan, Asynchronous task execution in networked control systems using decentralized event-triggering. Syst. Control Lett. **61**(9), 936–944 (2012)
76. M. Mazo, P. Tabuada, Decentralized event-triggered control over wireless sensor/actuator networks. IEEE Trans. Autom. Control **56**(10), 2456–2461 (2011)
77. M. Donkers, W. Heemels, Output-based event-triggered control with guaranteed L_∞ gain and improved and decentralized event-triggering. IEEE Trans. Autom. Control **57**(6), 1362–1376 (2012)
78. R. Olfati-Saber, R.M. Murray, Consensus problems in networks of agents with switching topology and time-delays. IEEE Trans. Autom. Control **49**(9), 1520–1533 (2004)
79. P.V. Teixeira, D.V. Dimarogonas, K.H. Johansson, J. Sousa, Multi-agent coordination with event-based communication, in *Proceedings of American Control Conference* (Baltimore, USA, 2010), pp. 824–829
80. T. Tang, Z. Liu, Z. Chen, Event-triggered formation control of multi-agent systems, in *Proceedings of 30th Chinese Control Conference* (Yantai, China, 2011), pp. 4783–4786
81. X. Chen, F. Hao, Event-triggered average consensus control for discrete-time multi-agent systems. IET Control Theory Appl. **6**(16), 2493–2498 (2012)
82. J. Hu, Second-order event-triggered multi-agent consensus control, in *Proceedings of 31st Chinese Control Conference* (Hefei, China, 2012), pp. 6339–6344
83. J. Hu, G. Chen, H.X. Li, Distributed event-triggered tracking control of second-order leader-follower multi-agent systems, in *Proceedings of 30th Chinese Control Conference* (Yantai, China, 2011), pp. 4819–4824
84. J. Hu, G. Chen, H.X. Li, Distributed event-triggered tracking control of leader-follower multi-agent systems with communication delays. Kybernetika **47**(4), 630–643 (2011)
85. D.V. Dimarogonas, E. Frazzoli, K.H. Johansson, Distributed event-triggered control for multi-agent systems. IEEE Trans. Autom. Control **57**(5), 1291–1297 (2012)
86. Z. Liu, Z. Chen, Z. Yuan, Event-triggered average-consensus of multi-agent systems with weighted and direct topology. J. Syst. Sci. Complex. **25**(5), 845–855 (2012)
87. Y. Fan, G. Feng, Y. Wang, C. Song, Distributed event-triggered control of multi-agent systems with combinational measurements. Automatica **49**(2), 671–675 (2012)
88. G.S. Seyboth, D.V. Dimarogonas, K.H. Johansson, Event-based broadcasting for multi-agent average consensus. Automatica **49**(1), 245–252 (2013)
89. X. Yin, D. Yue, S. Hu, Distributed event-triggered control of discrete-time heterogeneous multi-agent systems. J. Frankl. Inst. **350**(3), 651–669 (2013)

90. X. Yin, D. Yue, Event-triggered tracking control for heterogeneous multi-agent systems with Markov communication delays. J. Frankl. Inst. **350**(5), 1312–1334 (2013)
91. G. Welch, G. Bishop, An introduction to the Kalman filter, in Notes of ACM SIGGRAPH tutorial on the Kalman Filter (2001)
92. M. Miskowicz, Send-on-delta concept: an event-based data reporting strategy. Sensors **6**(1), 49–63 (2006)
93. A. Le, R. McCann, Event-based measurement updating Kalman filter in network control systems, in *Proceedings of IEEE Region 5 Technical Conference* (Fayetteville, USA, 2007), pp. 138–141
94. V.H. Nguyen, Y.S. Suh, Improving estimation performance in networked control systems applying the send-on-delta transmission method. Sensors **7**(10), 2128–2138 (2007)
95. Y.S. Suh, V.H. Nguyen, Y.S. Ro, Modified Kalman filter for networked monitoring systems employing a send-on-delta method. Automatica **43**(2), 332–338 (2007)
96. M. Miskowicz, Asymptotic effectiveness of the event-based sampling according to the integral criterion. Sensors **7**(1), 16–37 (2007)
97. V.H. Nguyen, Y.S. Suh, Networked estimation with an area-triggered transmission method. Sensors **8**(2), 897–909 (2008)
98. J. Sijs, M. Lazar, On event based state estimation, in *Hybrid Systems: Computation and Control* (Springer, Berlin, 2009), pp. 336–350
99. J. Sijs, M. Lazar, Event based state estimation with time synchronous updates. IEEE Trans. Autom. Control **57**(10), 2650–2655 (2012)
100. J. Wu, Q. Jia, K.H. Johansson, L. Shi, Event-based sensor data scheduling: trade-off between communication rate and estimation quality. IEEE Trans. Autom. Control **58**(4), 1041–1046 (2013)
101. L. Li, M. Lemmon, X. Wang, Event-triggered state estimation in vector linear processes, in *Proceedings of American Control Conference* (Baltimore, USA, 2010), pp. 2138–2143
102. M.G. Cea, G.C. Goodwin, Event based sampling in non-linear filtering. Control Eng. Pract. **20**(10), 963–971 (2012)
103. S. Hu, D. Yue, Event-based H_∞ filtering for networked system with communication delay. Signal Process. **92**(9), 2029–2039 (2012)
104. H. Li, Event-triggered state estimation for a class of delayed recurrent neural networks with sampled-data information. Abstr. Appl. Anal. **2012**, Article ID 731453, 21 p. (2012). https://doi.org/10.1155/2012/731453
105. S. Li, D. Sauter, B. Xu, Fault isolation filter for networked control system with event-triggered sampling scheme. Sensors **11**(1), 557–572 (2011)

Chapter 2
Feedback Stabilization of Networked Systems over Fading Channels

One primary issue of broad interest in the NCSs is the networked stabilization which aims to explore a fundamental limitation on the information constraints in order that the NCS can be stabilized. For a single-input system, such stabilization problems have been extensively studied under different information constraints. See [1–3] for data rate constraint, [4, 5] for packet drops, [6–8] for quantization, [9] for multiplicative stochastic noise, and [10] for signal-to-noise ratio constraint, etc. These studies have reached consensus regarding a unified fundamental limitation on the information constraints required for stabilization given in terms of the topological entropy of the open-loop plant, i.e., the logarithm of the absolute product of unstable poles for a discrete-time plant, or the sum of the unstable poles for a continuous-time plant. This builds an elegant bridge between the information requirement and the open-loop dynamics. In fact, as justified from various perspectives in the survey chapter [11], the topological entropy can be regarded as a measure of the degree of instability of a linear system. Bearing this in mind, it is quite intuitive that the larger topological entropy a system has, the more communication resource is needed for stabilization.

Promising progresses have been made toward the networked stabilization of multi-input systems as well. In particular, the idea of channel/controller codesign was initiated in [12] and followed by several other works, for instance, [13, 14] to address the multi-input networked stabilization. The essence therein is to employ the twist of channel resource allocation, i.e., assuming that the channel capacities can be allocated among different input channels subject to a total capacity constraint. By virtue of this additional design freedom, it has been shown under several different channel models, the minimum total capacity required for multi-input networked stabilization is still given by the topological entropy of the open-loop plant. Most recently, the work in [15] initiates the use of multiple-input and multiple-output (MIMO) communication in networked control leading to a certain coding/control codesign problem. This matchmaking of MIMO communication and MIMO control brings in new vitality to

Q. Liu et al., *Stochastic Control and Filtering over Constrained Communication Networks*, Studies in Systems, Decision and Control 178,
https://doi.org/10.1007/978-3-030-00157-5_2

the investigation of NCSs. A necessary and sufficient condition is obtained for the networked stabilizability given in terms of a majorization condition.

Motivated by these existing results, in this chapter, we study the multi-input networked stabilization over general fading channels subject to stationary SNR constraints. Each general fading channel is characterized by the cascade of a multiplicative stochastic noise and an additive Gaussian noise, which is a quite standard setup in communication theory. The work in this chapter can be considered as a unification of [12, 14], wherein the pure AWGN channels and pure fading channels have been addressed, respectively.

The first issue arising is how to handle the networked stabilization subject to both multiplicative noise and SNR-constrained additive noise. To this end, inspired by the work in [5], we attempt to build some sort of equivalence between a multiplicative stochastic noise and a deterministic gain followed by an additive noise subject to an instantaneous SNR constraint. It turns out that as far as mean-square stabilization is concerned, such an equivalence does hold and, thus, our previous experience in networked stabilization over SNR-constrained additive noise channels can be employed. Specifically, we exploit the idea of channel resource allocation which will turn the problem to a channel/controller codesign problem. By investigating the codesign problem, we arrive at a fundamental limitation on the communication channels which is again given by the topological entropy of the open-loop plant.

The remainder of this chapter is organized as follows: Sect. 2.1 formulates the problem to be studied in this chapter. Section 2.2 gives some preliminary knowledge for preparation. The main result is shown in Sect. 2.3. A numerical example is worked out for illustration in Sect. 2.4. Finally, the chapter is concluded in Sect. 2.5.

2.1 Problem Formulation

Consider a continuous-time system $[A|B]$ described by the state-space model as follows:

$$\dot{x}(t) = Ax(t) + Bu(t), \ x(0) = x_0, \tag{2.1}$$

where $x(t) \in \mathbb{R}^n$ is the system state, and $u(t) \in \mathbb{R}^m$ is the system input. Assume that A is unstable and $[A|B]$ is stabilizable. Also assume that the state $x(t)$ is available for feedback. For conventional control systems, the plant can directly employ the control signal $u(t) = Fx(t)$ with a properly designed gain matrix F to achieve closed-loop stability. However, when it comes to the NCSs wherein the information exchange between the controller and the plant is via a communication network, the feedback control signals from the remote controller may suffer from various distortions or information constraints due to the nonideal communication channels. This raises new challenges to control analysis and synthesis, even for the basic yet fundamental stabilization problem.

In this work, we consider the NSCs as in Fig. 2.1. The communication network is modeled as a set of parallel independent channels such that each control input is

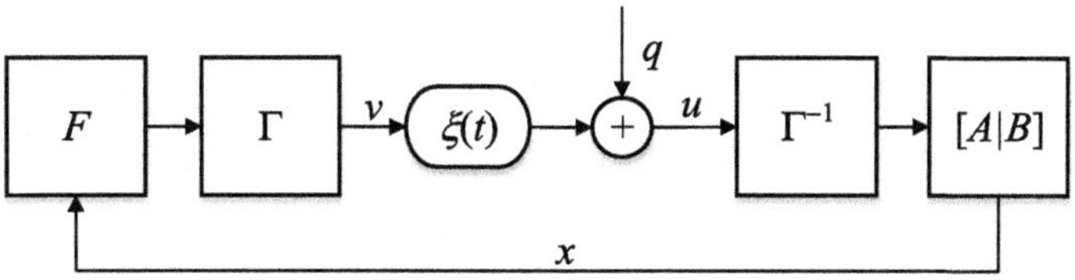

Fig. 2.1 Networked control systems

transmitted via a dedicated channel. Moreover, a pair of diagonal scaling matrices $\{\Gamma, \Gamma^{-1}\}$ with positive scalar entries are exploited at the transmitting and receiving sides, respectively, where

$$\Gamma = \mathrm{diag}\{\gamma_1, \gamma_2, ..., \gamma_m\}, \gamma_i > 0, \text{ for } i = 1, ..., m.$$

Such scaling matrices enable the possibility to adjust the transmission power in the different input channels. Similar techniques have been adopted in the literature as well [12, 16, 17].

Each channel in the communication network is modeled as a general fading channel, whose input–output relation is given by

$$u_i(t) = \xi_i(t)v_i(t) + q_i(t), \tag{2.2}$$

where the multiplicative noise $\xi_i(t)$ is a white Gaussian process with nonzero mean μ_i and autocovariance $\mathbb{E}[(\xi_i(t) - \mu_i)(\xi_i(t+\tau) - \mu_i)] = \sigma_i^2\delta(\tau)$, while the additive noise $q_i(t)$ is a white Gaussian process with zero mean and autocovariance $\mathbb{E}[q_i(t)q_i(t+\tau)] = p_i^2\delta(\tau)$. The input signal $v_i(t)$ is required to satisfy a stationary power constraint:

$$\mathbb{E}\{v_i^2\} < s_i^2,$$

which in fact places a stationary SNR constraint on the channel, namely, $\mathbb{E}\{v_i^2\}/p_i^2 < s_i^2/p_i^2$. By stacking all the channels together, we can describe the overall communication network as

$$u(t) = \xi(t)v(t) + q(t) \tag{2.3}$$

where

$$\xi(t) = \mathrm{diag}\{\xi_1(t), \xi_2(t), ..., \xi_m(t)\},$$
$$q(t) = \left[q_1(t)\ q_2(t)\ \ldots\ q_m(t)\right]'.$$

Denote

$$M = \text{diag}\{\mu_1, \mu_2, ..., \mu_m\},$$
$$\Sigma^2 = \text{diag}\{\sigma_1^2, \sigma_2^2, ..., \sigma_m^2\},$$
$$Q = \text{diag}\{p_1^2, p_2^2, ..., p_m^2\}.$$

For presentation compactness, we assume that $\mu_i = 1$, for $i = 1, \ldots, m$ hereinafter, without loss of generality.

The next question is how to characterize the information transmission capability of such fading channels. It has been well recognized that the Shannon capacity is, in general, not enough to characterize the information requirement for channels in a feedback system due to the causality constraint in the information processing in a feedback loop. How to resolve this issue from an information-theoretic point of view is intriguing and certainly requires more effort. Some inspiring attempts have been made recently in the literature, for instance, [18, 19]. Here, in this chapter, we come up with a capacity notion for the fading channels as follows:

$$\frac{1}{\mathfrak{C}_i} = \frac{1}{\mathfrak{C}_i^+} + \frac{1}{\mathfrak{C}_i^\times},$$

where

$$\mathfrak{C}_i^+ = \frac{1}{2}\frac{s_i^2}{p_i^2}, \quad \mathfrak{C}_i^\times = \frac{1}{2}\frac{1}{\sigma_i^2}.$$

The total channel capacity is then obtained by summing up all the channel capacities, i.e.,

$$\mathfrak{C} = \mathfrak{C}_1 + \mathfrak{C}_2 + \cdots + \mathfrak{C}_m. \tag{2.4}$$

Clearly, a high SNR constraint or low randomness of multiplicative noise will enlarge the capacity of a channel, which in turn, enhances the reliability of the information transmission through the channel.

The objective of this chapter is to find a fundamental limitation on the channel capacities $\mathfrak{C}_1, \mathfrak{C}_2, \ldots, \mathfrak{C}_m$, such that the NCS as in Fig. 2.1 can be stabilized via state feedback subject to the SNR constraint. What we mean by stabilization here is in fact in the mean-square sense, which will be clarified in the next section.

Remark 2.1 Although the capacity notion used in this chapter is somehow defined for the convenience of problem formulation and the aesthetics of result statement, it does agree with the existing understandings in the literature. Notice that the fading channel concerned can be regarded as a cascade of a multiplicative noise and an additive noise. When $\sigma_i = 0$, we have $\mathfrak{C}_i = \mathfrak{C}_i^+$, i.e., the capacity simply reduces to the capacity of an additive white Gaussian noise (AWGN) channel. Similarly, when $p_i = 0$, we have $\mathfrak{C}_i = \mathfrak{C}_i^\times$, i.e., the capacity reduces to the capacity of a multiplicative stochastic channel.

Remark 2.2 An intriguing observation is that the capacity notion (2.1) surprisingly admits the same form of the equivalent conductance of two cascaded electrical conductors. While the capacity of a channel measures the passage of information flow, the conductance of a conductor measures the passage of current flow. The same form of mathematical expression may suggest something more profound under the surface. Certain conceptual parallels may exist between the information flow in a feedback control loop and the current flow in an electrical circuit.

Before proceeding, recall that the topological entropy of a continuous-time linear system $\dot{x}(t) = Ax(t)$ is defined as $H(A) = \sum_{\Re(\lambda_i)>0} \lambda_i$, where λ_i are the eigenvalues of A and $\Re(\cdot)$ is the real part of a complex number.

2.2 Preliminary

In this section, some preliminary knowledge is presented for preparation, including the mean-square stability, Wonham decomposition, and $\mathcal{H}_2$ optimal complementary sensitivity.

2.2.1 *Mean-Square Stability*

Consider the NCS in Fig. 2.1. Substituting the input–output relation of the fading channel as in (2.3) into the system equation (2.1) yields the closed-loop dynamics:

$$\dot{x}(t) = (A + B\Gamma^{-1}\xi(t)\Gamma F)x(t) + B\Gamma^{-1}q(t). \tag{2.5}$$

Denote the state covariance by $X(t) = \mathbb{E}\{x(t)x'(t)\}$. Utilizing the Itö differential formula, the stochastic differential of $X(t)$ is given by

$$\begin{aligned}\dot{X}(t) &= (A + BMF)X(t) + X(t)(A + BMF)' \\ &\quad + B[\Sigma^2 \odot (FX(t)F')]B' + B\Gamma^{-1}Q\Gamma^{-1}B'.\end{aligned}$$

The closed-loop system is said to be mean-square stable(MSS) if for any given initial state $x(0)$, $X(t)$ is well defined for all t and $\lim_{t\to\infty} X(t) = \bar{X}$, where $\bar{X}$ is certain semi-positive definite matrix.

2.2.2 *Wonham Decomposition*

As indicated by its name, Wonham decomposition was initially proposed by [20] to solve the multi-input pole-placement problem. For later use, we briefly sketch the essence here. Given a stabilizable linear system $[A|B]$ with $A \in \mathbb{R}^{n\times n}$ and $B \in \mathbb{R}^{n\times m}$, we can carry out the controllable–uncontrollable decomposition with respect to the first column of B by a similarity transformation such that

$$[A|B] \simeq \left[\begin{bmatrix} A_1 & * \\ 0 & \tilde{A}_2 \end{bmatrix} \middle| \begin{bmatrix} b_1 & * \\ 0 & \tilde{B}_2 \end{bmatrix} \right].$$

Then, we proceed to do the controllable–uncontrollable decomposition to the system $[\tilde{A}_2|\tilde{B}_2]$ with respect to the first column of $\tilde{B}_2$. Continuing this process yields the following Wonham decomposition:

$$\left[\begin{bmatrix} A_1 & * & \dots & * \\ 0 & A_2 & \ddots & \vdots \\ \vdots & \ddots & \ddots & * \\ 0 & \dots & 0 & A_m \end{bmatrix} \middle| \begin{bmatrix} b_1 & * & \dots & * \\ 0 & b_2 & \ddots & \vdots \\ \vdots & \ddots & \ddots & * \\ 0 & \dots & 0 & b_m \end{bmatrix} \right]$$

that is equivalent to $[A|B]$, where $A_i \in \mathbb{R}^{n_i \times n_i}$ and $b_i \in \mathbb{R}^{n_i}$. Clearly, when $[A|B]$ is stabilizable, all the subsystems $[A_i|b_i]$, $i = 1, 2, \cdots, m$, are stabilizable.

2.2.3 Optimal Complementary Sensitivity

Consider the feedback system as in Fig. 2.1. Assume temporarily that the communication channels are ideal and the scaling pair is simply trivial, i.e., $\Gamma = I$. Then, the complementary sensitivity function at the plant input is given by

$$\mathcal{T}(s) = F(sI - A - BF)^{-1}B.$$

As shown in many existing works [9, 10, 14] as well as later developments in this work, the feedback stabilization in the presence of additive noise or multiplicative noise is closely related to the $\mathcal{H}_2$ optimal $\mathcal{T}(s)$. For preparation, the following lemma gives a solution to $\mathcal{H}_2$ optimal $\mathcal{T}(s)$.

Lemma 2.2.1 ([21]) *There holds*

$$\inf_{F: A+BF \text{ is stable}} \|\mathcal{T}(s)\|_2 = [2H(A)]^{\frac{1}{2}}.$$

Moreover, when A has no eigenvalues on the imaginary axis, the infimum can be achieved by the optimal state feedback gain $F = -B'X$, where X is the stabilizing solution to the algebraic Riccati equation

$$A'X + XA - XBB'X = 0.$$

2.3 Main Results

In this section, we are dedicated to establishing the minimum channel capacity required for feedback stabilization with SNR constraints over general fading channels.

At the very beginning, let us introduce an auxiliary networked control system as in Fig. 2.2, where the fading channel is replaced by a constant gain M plus an additive white Gaussian noise with mean zero and power spectral density $P_d = \Sigma^2 \odot P_{\upsilon^a} + Q$. By doing so, we are able to establish an equivalence between the MSS of the NCS as in Fig. 2.1 and the internal stability of the auxiliary system as in Fig. 2.2 subject to a certain SNR constraint. As such, we can focus on studying the auxiliary system which is easier to handle.

A more precise statement of the above-mentioned equivalence is presented in the following Lemma:

Lemma 2.3.1 *Consider the two NCSs shown in Figs. 2.1 and 2.2. The following statements hold:*

(1) the NCS of Fig. 2.1 is mean-square stable if and only if the NCS of Fig. 2.2 is internally stable and there exists a finite positive-semidefinite choice $P_d \geq 0$ *such that* $P_d = \Sigma^2 \odot P_{\upsilon^a} + Q$ *holds, where* P_{υ^a} *denotes the steady-state power spectral density of the signal* υ^a *in Fig. 2.2;*
(2) if NCS of Fig. 2.2 is internally stable and there exists $P_d \geq 0$ *such that* $P_d = \Sigma^2 \odot P_{\upsilon^a} + Q$*, then, for the same choice of* P_d*, the steady-state power spectral density of the signal* υ^a *in Fig. 2.2 equals that of the signal* υ *in Fig. 2.1, i.e.,* $P_\upsilon = P_{\upsilon^a}$.

This lemma is the continuous-time multichannel counterpart of the theorems in [5, 22]. The proof is omitted here for brevity.

The next theorem gives a necessary and sufficient condition for the mean-square stabilization of the NCSs over general fading channels with SNR constraints.

Theorem 2.3.2 *The NCS shown in Fig. 2.1 over general fading channels with SNR constraints can be mean-square stabilized, if and only if* $\mathfrak{C} > H(A)$.

Proof Necessity: In view of Lemma 2.3.1, when the system in Fig. 2.1 is MSS, the stationary power spectral density $P_\upsilon = P_{\upsilon^a}$, i.e., the power of signal υ in Fig. 2.1 is equal to the power of signal υ^a in Fig. 2.2.

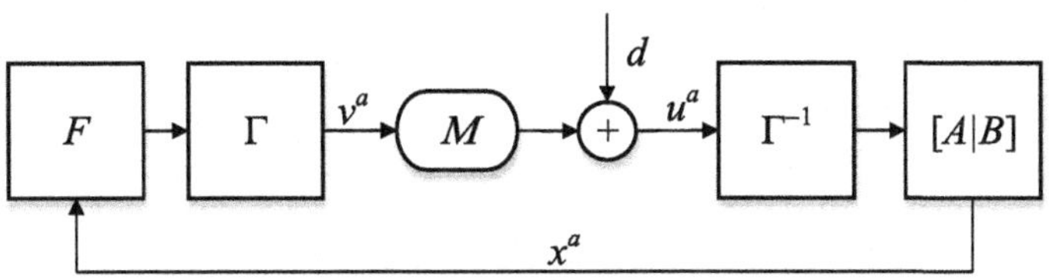

Fig. 2.2 Auxiliary networked control systems

In the auxiliary system, denote the complementary sensitivity function as

$$T(s) = \Gamma F(sI - A - BF)^{-1}B\Gamma^{-1}.$$

It is not hard to verify that the stationary power spectral density of the signal v^a is given by

$$P_{v^a} = \frac{1}{2\pi}\int_{-\infty}^{\infty} T(j\omega)P_d T(j\omega)^* d\omega. \tag{2.6}$$

Pre-multiplying and post-multiplying both sides of (2.6) by $P_d^{-\frac{1}{2}}$, it yields

$$\frac{\mathbb{E}[(v_i^a)^2]}{\sigma_i^2\mathbb{E}[(v_i^a)^2]+p_i^2} = \frac{1}{2\pi}\int_{-\infty}^{\infty}\{P_d^{-\frac{1}{2}}T(j\omega)P_d T(j\omega)^* P_d^{-\frac{1}{2}}\}_{ii}d\omega,$$

where v_i^a is the ith element of the vector v^a and $\{.\}_{ii}$ represents the ith diagonal element of a matrix. In view of Lemma 2.2.1, we have

$$\sum_{i=1}^{m}\frac{\mathbb{E}[(v_i^a)^2]}{\sigma_i^2\mathbb{E}[(v_i^a)^2]+p_i^2} = \|P_d^{-\frac{1}{2}}T(s)P_d^{\frac{1}{2}}\|_2^2 \geq 2H(A),$$

where the first equality follows from the diagonality of the matrix P_d. Moreover, under the SNR constraints, the stationary power level $\mathbb{E}[v_i^2] = \mathbb{E}[(v_i^a)^2] < s_i^2$. It follows that

$$H(A) < \frac{1}{2}\sum_{i=1}^{m} 1\Big/\left(\sigma_i^2 + \frac{p_i^2}{s_i^2}\right) = \mathfrak{C}_1 + \cdots + \mathfrak{C}_m = \mathfrak{C},$$

which ends the necessity proof.

Sufficiency: The sufficiency part is shown through a channel/controller codesign. Without loss of generality, $[A|B]$ is assumed to be in the Wonham decomposition form. In view of Lemma 2.2.1, for each stabilizable subsystem $[A_i|b_i]$, a stabilizing feedback gain f_i can be designed such that $\|T_i(s)\|_2^2 = 2H(A_i)$, where

$$T_i(s) = f_i(sI - A_i - b_i f_i)^{-1}b_i.$$

We then choose

$$F = \text{diag}\{f_1, f_2, ..., f_m\}$$

so as to make the closed-loop system in Fig. 2.2 internally stable. Since $\mathfrak{C} > H(A)$, there always exists a positive scalar ϵ satisfying $\mathfrak{C} = H(A) + \epsilon$. As a result, the resource allocation can be done by letting $\mathfrak{C}_i = H(A_i) + \epsilon/m$. Now, choose the scaling matrix $\Gamma = D^{-1}$ with $D = \text{diag}\{1, \eta, ..., \eta^{m-1}\}$, where η is a small positive number. Also, denote $S = \text{diag}\ \{I_{n_1}, \eta I_{n_2}, ..., \eta^{m-1}I_{n_m}\}$. Then

$$T(s) = \Gamma F(sI - A - BF)^{-1} B \Gamma^{-1}$$
$$= \mathcal{F}(sI - \mathcal{A} - \mathcal{B}\mathcal{F})^{-1}\mathcal{B},$$

where

$$\mathcal{F} = D^{-1} F S = F,$$

$$\mathcal{A} = S^{-1} A S = \begin{bmatrix} A_1 & o(\eta) & \dots & o(\eta) \\ 0 & A_2 & \ddots & \vdots \\ \vdots & \ddots & \ddots & o(\eta) \\ 0 & \dots & 0 & A_k \end{bmatrix},$$

$$\mathcal{B} = S^{-1} B D = \begin{bmatrix} b_1 & o(\eta) & \dots & o(\eta) \\ 0 & b_2 & \ddots & \vdots \\ \vdots & \ddots & \ddots & o(\eta) \\ 0 & \dots & 0 & b_k \end{bmatrix},$$

and $o(\eta)/\eta$ approaches to a finite constant as $\eta \to 0$. Therefore,

$$T(s) = \ \mathrm{diag}\{T_1(s), ..., T_m(s)\} + o(\eta).$$

Let the auxiliary noise power spectral density be $P_d = \Sigma^2 \odot P_{v^a} + Q$. We can have the following expression via a series of matrix manipulations

$$\frac{\mathbb{E}[(v_i^a)^2]}{\sigma_i^2 \mathbb{E}[(v_i^a)^2] + p_i^2} = \frac{1}{2\pi} \int_{-\infty}^{\infty} \{P_d^{-\frac{1}{2}} T(j\omega) P_d T(j\omega)^* P_d^{-\frac{1}{2}}\}_{ii} d\omega \ = \|T_i(s)\|_2^2 + o(\eta).$$

Since

$$\|T_i(s)\|_2^2 = 2H(A_i) = 2\mathfrak{C}_i - \epsilon/m,$$

it follows that when η takes a sufficiently small value, there holds

$$\frac{\mathbb{E}[(v_i^a)^2]}{\sigma_i^2 \mathbb{E}[(v_i^a)^2] + p_i^2} < 1 \Big/ \left(\sigma_i^2 + \frac{p_i^2}{s_i^2}\right),$$

which further yields

$$\mathbb{E}[(v_i^a)^2] < s_i^2, \ \text{ for } \ i = 1, 2, ..., m.$$

Now, in view of Lemma 2.3.1, one can conclude that the NCS as in Fig. 2.1 is MSS, and there holds $\mathbb{E}[v_i^2] = \mathbb{E}[(v_i^a)^2] < s_i^2$, i.e., the SNR constraints are satisfied. The sufficiency proof ends now.

Table 2.1 Channel/controller codesign procedure

Procedure	
Step 1	Choose a proper scaling matrix Γ with sufficient small η
Step 2	Allocate the total channel capacity $\mathfrak{C}$ among all the input subchannels such that $\mathfrak{C}_i > H(A_i)$
Step 3	Design the control gains f_i by solving a set of typical $\mathcal{H}_2$ optimal control problems

The above theorem reveals that the infimum of the total channel capacity needed for stabilizing an NCS with transmission SNR constraints is given by the topological entropy of the open-loop plant. Moreover, the constructive sufficiency proof also gives us a guideline on how to explicitly allocate the resource as well as design the controller. Precisely, the channel/controller codesign procedure is trifold as shown in Table 2.1.

2.4 An Illustrative Example

In this section, the main result presented before is demonstrated by a numerical example.

Consider an unstable system $[A|B]$ with

$$A = \begin{bmatrix} 2 & 0 & 0 \\ 0 & 1 & 0 \\ 0 & 0 & 1 \end{bmatrix}, \quad B = [B_1 \ B_2] = \begin{bmatrix} 1 & 0 \\ 1 & 1 \\ 0 & 1 \end{bmatrix}.$$

It can be seen that $[A|B]$ is stabilizable. Nonetheless, $[A|B]$ cannot be transferred into a stabilizable single-input system by just combining the input signals from two channels. In other words, $[A|b\,]$ is not stabilizable for any $b \in \text{span}\{B_1, B_2\}$, since the matrix $[\lambda I - A|b\,]$ loses row rank when $\lambda = 1$. Note that $[A|B]$ is already in the Wonham decomposition form with the following two subsystems:

$$[A_1|\, b_1] \text{ and } [A_2|\, b_2],$$

where

$$A_1 = \begin{bmatrix} 2 & 0 \\ 0 & 1 \end{bmatrix}, \quad b_1 = \begin{bmatrix} 1 \\ 1 \end{bmatrix}, \quad A_2 = 1, \ \ b_2 = 1.$$

The topological entropy of the subsystems are given by $H(A_1) = 3$ and $H(A_2) = 1$, respectively. As for the general fading channels, assume that the multiplicative noise has mean I and the additive noise has the power spectral density $Q = \text{diag}\{0.22,\ 0.19\}$.

According to Theorem 1, the NCS is mean-square stabilizable under SNR constraints if and only if the total channel capacity $\mathfrak{C} > H(A_1) + H(A_2) = 4$. In what follows, we illustrate how exactly the resource allocation is done. Assume that the total channel capacity is given by $4 + 2 \times 10^{-3}$. Then, we allocate the channel capacity between two input channels with $\mathfrak{C}_1 = 3 + 10^{-3} > H(A_1)$ and $\mathfrak{C}_2 = 1 + 10^{-3} > H(A_2)$, respectively. Such resource allocation can be realized by adjusting the stationary transfer power s_i^2 and the power spectral density of the multiplicative noise σ_i^2. Specifically, let

$$\Sigma^2 = \text{diag}\{9,\ 5\},\ \ s_1^2 = 4,\ \ s_2^2 = 0.64.$$

In order to find the feedback gains $f_i, i = 1, 2$, we solve the $\mathcal{H}_2$ optimal controller as in Lemma 2.2.1 leading to

$$f_1 = \begin{bmatrix} -12 & 6 \end{bmatrix},\ \ f_2 = -2.$$

It follows that

$$F = \begin{bmatrix} f_1 & 0 \\ 0 & f_2 \end{bmatrix}.$$

Further, design the scaling matrix Γ as

$$\Gamma = \text{diag}\{1, 10^{-3}\}.$$

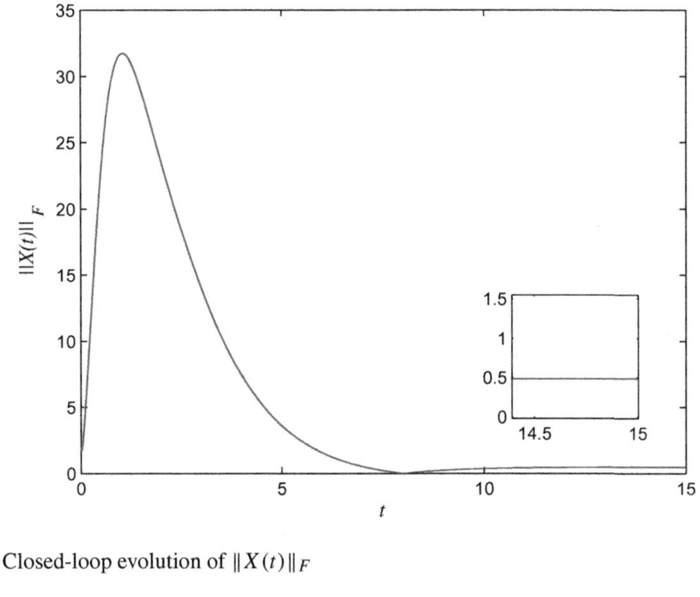

Fig. 2.3 Closed-loop evolution of $\|X(t)\|_F$

In this way, we finish the channel-controller codesign. By employing the numerical simulation tool MATLAB, we can see from Fig. 2.3 that the Frobenius norm of the state covariance converges to a finite constant and, thus, the closed-loop system is MSS. Moreover, further computation yields that

$$\mathbb{E}\{v_1^2\} = 3.9960 < s_1^2 = 4,$$
$$\mathbb{E}\{v_2^2\} = 0.6389 < s_2^2 = 0.64,$$

which indicates that the SNR constraints are indeed fulfilled.

2.5 Conclusions

In this chapter, we investigate the networked stabilization over general fading channels with stationary SNR constraints. Each general fading channel is modeled as the cascade of a stochastic multiplicative noise and an additive white Gaussian noise. A capacity notion is introduced to measure the information transmission capability of such a channel. We wish to find the minimal total channel capacity rendering networked stabilization possible. To this end, one auxiliary feedback system is proposed, in which the multiplicative noise in a general fading channel is replaced by a constant gain followed by an SNR-constrained additive noise. By doing so, we are able to translate the original mean-square stabilization problem to the stabilization of the auxiliary system subject to certain SNR constraints. By resorting to the channel-controller codesign idea, we obtain a necessary and sufficient condition on the information requirement for mean-square stabilization given in terms of the topological entropy of the open-loop plant. Also, a numerical example has been given to verify the effectiveness of the proposed results.

References

1. J. Baillieul, P. Antsaklis, Control and communication challenges in networked real-time systems. Proc. IEEE **95**(1), 9–28 (2007)
2. G.N. Nair, R.J. Evans, Exponential stabilisability of finite-dimensional linear systems with limited data rates. Automatica **39**(4), 585–593 (2003)
3. S. Tatikonda, S. Mitter, Control under communication constraints. IEEE Trans. Autom. Control **49**, 1056–1068 (2004)
4. K. You, L. Xie, Minimum data rate for mean square stabilization of discrete LTI systems over lossy channels. IEEE Trans. Autom. Control **55**(10), 2373–2378 (2010)
5. E.I. Silva, M.A. Solis, Control of LTI plants over erasure channels. Automatica **47**(8), 1729–1736 (2011)
6. N. Elia, S. Mitter, Stabilization of linear systems with limited information. IEEE Trans. Autom. Control **46**(9), 1384–1400 (2001)
7. M. Fu, L. Xie, The sector bound approach to quantized feedback control. IEEE Trans. Autom. Control **50**(11), 1698–1711 (2005)

8. Z. Wang, B. Shen, H. Shu, G. Wei, Quantized control for nonlinear stochastic time-delay systems with missing measurements. IEEE Trans. Autom. Control **57**(6), 1431–1444 (2012)
9. N. Elia, Remote stabilization over fading channels. Syst. Control Lett. **54**(3), 237–249 (2005)
10. J.H. Braslavsky, R.H. Middleton, J.S. Freudenberg, Feedback stabilization over signal-to-noise ratio constrained channels. IEEE Trans. Autom. Control **52**(8), 1391–1403 (2007)
11. L. Qiu, Quantify the unstable, in *Proceedings of the 19th International Symposium on Mathematical Theory of Networks and Systems-MTNS*, vol. 5, no. 9 (2010)
12. L. Qiu, G. Gu, W. Chen, Stabilization of networked multi-input systems with channel resource allocation. IEEE Trans. Autom. Control **58**(3), 554–568 (2013)
13. W. Chen, L. Qiu, Stabilization of networked control systems with multirate sampling. Automatica **49**(6), 1528–1537 (2013)
14. N. Xiao, L. Xie, L. Qiu, Feedback stabilization of discrete-time networked systems over fading channels. IEEE Trans. Autom. Control **57**(9), 2176–2189 (2012)
15. W. Chen, S. Wang, L. Qiu, When MIMO control meets MIMO communication: a majorization condition for networked stabilizability (2014), arXiv preprint arXiv:1408.3500
16. J.S. Freudenberg, R.H. Middleton, J.H. Braslavsky, Stabilization with disturbance attenuation over a Gaussian channel, in *Proceedings of the 46st IEEE Conference Decision Control*, (New Orleans, 2007), pp. 3958–3963
17. F. Vargas, J. Chen, E.I. Silva, On stabilizability of MIMO systems over parallel noisy channels, in *Proceedings of the 53rd IEEE Conference Decision Control*, (Los Angeles, 2014), pp. 6074–6079
18. A. Sahai, S. Mitter, The necessity and sufficiency of anytime capacity for stabilization of a linear system over a noisy communication link-part I: scalar systems. IEEE Trans. Inf. Theory **52**, 3369–3395 (2006)
19. G.N. Nair, A nonstochastic information theory for communication and state estimation. IEEE Trans. Autom. Control **58**, 1497–1510 (2013)
20. W. Wonham, *Linear Multivariable Control: A Geometrix Approach*, 3rd edn. (Springer, 1985)
21. W. Chen, Topological entropy of continuous-time linear systems, M.Phil. Thesis, Hong Kong University of Science and Technology, 2010
22. A.I. Maass, E.I. Silva, Performance limits in the control of single-input linear time-invariant plants over fading channels. IET Control Theory Appl. **8**(14), 1384–1395 (2014)

Chapter 3
Event-Based H_∞ Consensus Control of Multi-agent Systems

During the past decade, the coordination problems of multi-agent systems have been a research focus attracting an increasing interest due primarily to their practical application insights in a variety of realms such as satellite formation control [1], collective behavior of flocking [2], attitude alignment among spacecraft [3], distributed estimation [4], and automated highway systems [5]. A critical issue in coordinated control problems is to design a distributed control protocol for communication behaviors and controller actuation based on the shared information (including graph topologies, real-time states of adjacent nodes and common control algorithms) in order to ensure that all the agents reach an agreement or collectively perform certain actions. Many important results have recently been reported on the cooperative control problems for multi-agent systems, see, e.g., [6–15] and the references therein.

Consider the practical situations where the states of the multi-agent systems are subject to real-time changes/variations/fluctuations such as time-varying temperature and mutative working conditions. In this circumstance, the evolution of the dynamics of local agents is inevitably dependent on the time. For such time-varying systems, in response to the changes in the environment, the intelligent agents should adopt the *time-varying* cooperative control strategies so as to better reflect the reality. A literature search has shown that several methodologies have recently been developed for time-varying systems and a great number of results have been available for the general control and filtering problems. These methodologies include, but are not limited to, the recursive linear matrix inequality (RLMI) technique [16] and the backward recursive Riccati difference equation (RDE) approach [17]. Unfortunately, up to now, the corresponding results on the coordination problems for time-varying multi-agent systems have been really scattered, and the first motivation of this chapter is to shorten such a gap.

Owing to the recent advances in digital technologies, embedded microprocessors, which are responsible for communication between adjacent agents, have been becoming indispensable components of multi-agent systems. In an *ideal* world, the

Q. Liu et al., *Stochastic Control and Filtering over Constrained Communication Networks*, Studies in Systems, Decision and Control 178,
https://doi.org/10.1007/978-3-030-00157-5_3

communication bandwidth is assumed to be unlimited and the traditionally periodically triggered communication won't bring any concern in terms of the network load. Such an assumption, however, is not true in some applications subject to certain resource constraints such as limited network bandwidth. For example, the frequent signal transmissions might give rise to network-related adverse phenomena such as communication delays and packet losses. In this case, a novel sporadic scheduling (called the event-based approach) seems to be more preferable, where the pre-described executions are triggered if and only if some "interesting" events occur. With appropriately developed triggering events, it is predictable that both the reduction of bandwidth occupation and the desired properties of closed-loop system (e.g., stability and convergence) can be guaranteed [18].

Up to now, many event-based schemes have been available in the literature for continuous- or discrete-time systems based on the input-to-state stable (ISS) theory, see for instance [19] and the references therein. Such sporadic event-based schemes have been applied in [20] for distributed network control systems (NCSs) with packet loss and transmission delays. Event-based control problems have also been addressed for multi-agent systems, see, e.g., [21–25]. Specifically, in [22], both centralized and decentralized event-based control strategies have been proposed for a group of single-integrator multi-agents in order to reach an agreement according to a fixed undirected network topology. The event-based tracking control problem has been investigated in [21] for leader–follower multi-agent systems with and without communication delays, where the convergence analysis has been provided. Several linear matrix inequality (LMI) conditions have been reported in [23] for event-based control problem of discrete-time heterogeneous multi-agent systems. Additionally, in [25], the distributed event-based methods have been combined with an iterative algorithm to render the implementation more practical. A novel event-based strategy, which is independent of the real-time state of neighbors, has been examined in [24] for both single and double-integrator agents such that the continuous monitoring is no longer required. As for general agent dynamics, in [26], the authors have intensively investigated the consensus of multi-agent systems when the individual full state is available for its neighbors. Furthermore, in [27, 28], the event-based consensus problems have been thoroughly studied for general linear or nonlinear system dynamics by assuming that each agent is passive.

It should be pointed out, despite the recent surge of research attention on the event-based schemes for multi-agent systems, several challenges still remain. First, most available results have been concerned with single- or double-integrator *time-invariant* models *without* any external disturbances. Unfortunately, in real-world applications, the behaviors of local agents are usually complicated especially when they suffer from various stochastic disturbances and communication-induced noises. It is of vital importance to suppress the influence from the external noise disturbances for general multi-agent systems through analyzing and synthesizing the cooperative control schemes, for which the H_∞ disturbance rejection attenuation would be a suitable performance index. So far, some preliminary results [29, 30] have been reported on the H_∞ consensus control problem with the assumption that real-time information of adjacent nodes' *full states* is available in order for the agents to share their local information at every sampling instant. Such an assumption is somewhat restrictive in practice since the state of agents may not be easily available in some occasions. Also, the real-time full state feedback would increase the communication frequency, which is undesirable. To resolve the problem, a natural idea is to develop an appropriate event-based control scheme by using *accessible relative measurement outputs* instead of the full states, and this motivates us to investigate the H_∞ consensus control problem.

Summarizing the above discussions, in this chapter, we aim to deal with the H_∞ *consensus* problems for a class of linear *time-varying* multi-agent systems with external disturbances by using *relative output feedback*. An event-based scheme is proposed to distributively calculate event times based on a predetermined data transmission condition in order to reduce the communication burden among agents. Moreover, we would like to derive sufficient conditions under which the consensus error is bounded in an H_∞ sense over a finite horizon. The corresponding time-varying estimation-based output feedback controller is designed via optimizing an H_2 performance index. The main contributions can be highlighted as follows: *(1) the discrete-time intelligent agents with general dynamics are under consideration which cover the frequently investigated integrator models as special cases; (2) the transient behaviors are studied in order to reflect the time-varying nature of the addressed multi-agent systems; (3) a novel event-based control protocol is first proposed for the time-varying multi-agent systems so as to achieve the prespecified* H_∞ *constraints over a finite horizon* $[0, T]$; *and (4) different from the existing literature, relative measurements between adjacent agents are utilized for the event-based feedback control.*

3.1 Problem Formulation

In this section, we introduce some preliminaries related to distributed control of multi-agent systems and then describe the problem setup.

3.1.1 Graph Topologies

The communication topology of the system is described by a fixed undirected graph $\mathcal{G} = (\mathcal{V}, \mathcal{E}, \mathcal{A})$ of order N with the set of nodes $\mathcal{V} =\{1, 2, \ldots, N\}$, the set of edges $\mathcal{E} = \mathcal{V} \times \mathcal{V}$, and the weighted adjacency matrix $\mathcal{A} =[a_{ij}]$. The weighted adjacency matrix of the graph is a matrix with nonnegative elements a_{ij} satisfying the property $a_{ij} > 0 \Longleftrightarrow (i, j) \in \mathcal{E}$, which means if there is an edge between nodes i and j, then they are called adjacent. Note that for an undirected graph $\mathcal{G}$, $\mathcal{A}$ is a symmetric $N \times N$ matrix given by $a_{ij} = 1$ (if nodes i and j are adjacent) and $a_{ij} = 0$ (otherwise). The graph $\mathcal{G}$ is assumed to be connected where there is a path between any of two nodes. The neighbors of node i are denoted by $\mathcal{N}_i \triangleq \{j|(i, j) \in \mathcal{E}\}$. The degree D is a diagonal matrix with elements d_i defined as the number of its adjacent vertices. The Laplacian of undirected graph $\mathcal{G}$ is a symmetric positive semidefinite matrix $H = D - A$.

3.1.2 Multi-agent Systems

Consider a multi-agent system with N identical agents, labeled by $1, 2, \ldots, N$, respectively. The dynamics of agent i is governed by the following discrete time-varying systems:

$$\begin{cases} x_i(k+1) = A_k x_i(k) + B_k u_i(k) + D_k w_i(k), \\ y_i(k) = C_k x_i(k) + E_k v_i(k), \\ z_i(k) = M_k x_i(k), \end{cases} \tag{3.1}$$

where $x_i(k) \in \mathbb{R}^{n_x}$ is the system state, $w_i(k) \in \mathbb{R}^{n_w}$ and $v_i(k) \in \mathbb{R}^{n_v}$ are the external disturbances belonging to $\mathcal{L}_2[0, T]$, $y_i(k)$ is the measurement output, $z_i(k)$ is the controlled output, and $u_i(k)$ is a sequence of control inputs. Note that the system (3.1) under consideration is quite general which includes the single- and second-order integrators as special cases.

Owing to the time-varying manner of the system and the influence from external disturbances, it is difficult for the multi-agents to achieve asymptotic and accurate consensus. Therefore, in this chapter, the finite-horizon H_∞ consensus problem is taken into account to alleviate the interferences from the time-varying parameters and the external disturbances on the desired agreement among the agents.

Definition 3.1.1 Let a disturbance attenuation level $\gamma > 0$ and a positive definite matrix $W = W^T > 0$ be given. The multi-agent system (3.1) with a prefixed connected topology is said to satisfy the H_∞ consensus performance constraint over the finite horizon $[0, T]$ if the following inequality holds:

$$\sum_{i=1}^{N} \|\bar{z}_i(k)\|^2_{[0,T]} < \gamma^2 \sum_{i=1}^{N} \left\{ \|w_i(k)\|^2_{[0,T]} + \|v_i(k)\|^2_{[0,T]} + \bar{x}_i^T(0) W \bar{x}_i(0) \right\}, \tag{3.2}$$

where $\bar{z}_i(k) = z_i(k) - \frac{1}{N}\sum_{j=1}^{N} z_i(k)$, $\|\bar{z}_i(k)\|^2_{[0,T]} = \sum_{k=0}^{T} \|\bar{z}_i(k)\|^2$, and $\bar{x}_i(0) = x_i(0) - \frac{1}{N}\sum_{j=1}^{N} x_i(0)$.

Remark 3.1 The asymptotic (steady-state) H_∞ consensus problem has been dealt with in [30] for time-invariant system over an infinite horizon. To capture the behaviors of time-varying systems addressed in this chapter, it makes more sense to study the finite-horizon (transient) H_∞ consensus control problem with the hope to attenuate the effects from external disturbances over a specific time period.

3.1.3 Cooperative Estimators Design

Consider the situation where each agent has access to the measurements relative to its adjacent agents rather than the local measurements, which means the measurements $y_i(k)$ $(i = 1, 2, \ldots, N)$ cannot be obtained directly. In the following, a distributed estimator-type consensus protocol is proposed based on relative output measurements, which utilizes the state estimation information (for the controller design) and the output measurement information (for the estimator design).

The relative measurement of adjacent agents with respect to agent i is defined by

$$\zeta_i(k) = \sum_{j\in\mathcal{N}_i} a_{ij}(y_j(k) - y_i(k)), \quad \zeta_i(k) \in \mathbb{R}^{n_y} \tag{3.3}$$

and the relative full state is defined by

$$\xi_i^*(k) = \sum_{j\in\mathcal{N}_i} a_{ij}(x_j(k) - x_i(k)), \quad \xi_i^*(k) \in \mathbb{R}^{n_x}. \tag{3.4}$$

Note that the relative full state is unavailable but could be estimated. As such, an estimator-type consensus protocol is proposed as

$$\begin{cases} \xi_i(k+1) = A_k\xi_i(k) + B_k \sum_{j\in\mathcal{N}_i} a_{ij}(u_j(k) - u_i(k)) \\ \qquad\qquad + L_k(\zeta_i(k) - C_k\xi_i(k)) \\ u_i(k) = K_k\xi_i(k), \end{cases} \tag{3.5}$$

where $\xi_i(k) \in \mathbb{R}^{n_x}$ is an estimate for the variable $\xi_i^*(k)$ and L_k (K_k) are the estimator (controller) parameters to be determined with appropriate dimensions. The control term $\sum_{j\in\mathcal{N}_i}(u_j(k) - u_i(k))$ in (3.5) requires the information exchanges between each agent and its neighbors, which renders the protocol (3.5) distributed and convenient for implementation in practical applications.

3.1.4 Event-Based Mechanism

Thanks to its capabilities to reduce the information exchange frequency, the event-based mechanism has proven to be suitable for distributed control of multi-agent systems subject to limited network resources.

For the purpose of introducing the event-based scheduling, we first denote the triggering instant sequence of agent i by $s_0^i = 0 < s_1^i < s_2^i < \cdots$ and then define $\xi_i^t(k) = \xi_i(s_m^i)$ for $k \in [s_m^i, s_{m+1}^i)$ with the superscript "t" indicating triggering. To this end, the event-based estimator-type consensus protocol (3.5) can be rewritten as follows with a little abuse of the notation $\xi_i(k)$:

$$\begin{cases} \xi_i(k+1) = A_k\xi_i(k) + B_k \sum_{j\in\mathcal{N}_i} a_{ij}(u^t_j(k) - u^t_i(k)) \\ \qquad\qquad + L_k(\zeta_i(k) - C_k\xi_i(k)) \\ u^t_i(k) = K_k\xi^t_i(k) \end{cases} \tag{3.6}$$

and the dynamics of the closed-loop system can be rewritten as

$$\begin{cases} x_i(k+1) = A_k x_i(k) + B_k u^t_i(k) + D_k w_i(k), \\ y_i(k) = C_k x_i(k) + E_k v_i(k), \\ z_i(k) = M_k x_i(k). \end{cases} \tag{3.7}$$

Moreover, let σ be a given positive scalar and the distributed triggering function $f_i(k, \xi_i(k), \xi^t_i(k), \sigma)$ be given by

$$f_i(k) = (\xi_i(k) - \xi^t_i(k))^T(\xi_i(k) - \xi^t_i(k)) - \sigma\xi^T_i(k)\xi_i(k), \tag{3.8}$$

which takes values in $\mathbb{R}$ for each agent. The control execution is triggered as long as the inequality $f_i(.) > 0$ is satisfied, i.e.,

$$(\xi_i(k) - \xi^t_i(k))^T(\xi_i(k) - \xi^t_i(k)) > \sigma\xi^T_i(k)\xi_i(k). \tag{3.9}$$

Therefore, the next triggering instant is determined iteratively by

$$s^i_{m+1} = \min\{k \in \mathbb{N} | k > s^i_m,\ f_i(k, \xi_i(k), \xi^t_i(k), \sigma) > 0\}. \tag{3.10}$$

Remark 3.2 Notice that the triggering function in [22] makes the implicit assumptions that each agent should be aware of the real-time full states of its neighbors, and thereby causing some applicative difficulties for certain occasions when the neighbor's states could not be directly observed. However, in this chapter, only relative output measurements are accessible for each agent, therefore, the event generator (3.10) should be under serious reconsideration to make it implementable in the sense of distributed framework.

Remark 3.3 From the event-based protocol (3.6), it can be seen that the estimator constructed for each agent updates the signals $\sum_{j\in\mathcal{N}_i}(u^t_j(k) - u^t_i(k))$ at event-triggering instants for both the adjacent agents and itself, while the control input signals are corrected only when the agent triggers an event. On the other hand, once (3.9) is satisfied, a new event is triggered to correct the difference of the estimation $\xi_i(k)$ (local knowledge) and the estimation at event times $\xi^t_i(k)$ (shared knowledge between adjacent nodes). According to the triggering rules (3.10), at triggering instants, we have $\xi^t_i(k) = \xi_i(k)$ which indicates that $f_i(k, \xi_i(k), \xi^t_i(k), \sigma) \le 0$ would never be violated during the system process. Moreover, the scalar σ regulates the triggering frequency. Obviously, more events would be triggered if such a scalar decreases. Particularly, when $\sigma = 0$, the event-based control approach reduces to the classical clock-driven control one.

To facilitate the subsequent formulation, the corresponding estimation error and the control input error (between actual and ideal input signals) are defined, respectively, by

$$e_i^s(k) = \xi_i(k) - \xi_i^*(k), \quad e_i^t(k) = \xi_i^t(k) - \xi_i^*(k).$$

For notational presentation convenience, here we denote

$$\begin{aligned} x(k) &= \mathrm{col}_N\{x_i(k)\},\ \xi^*(k) = \mathrm{col}_N\{\xi_i^*(k)\},\ \xi(k) = \mathrm{col}_N\{\xi_i(k)\} \\ v(k) &= \mathrm{col}_N\{v_i(k)\},\ w(k) = \mathrm{col}_N\{w_i(k)\},\ e^s(k) = \mathrm{col}_N\{e_i^s(k)\} \\ e^t(k) &= \mathrm{col}_N\{e_i^t(k)\},\ u^t(k) = \mathrm{col}_N\{u_i^t(k)\},\ u(k) = \mathrm{col}_N\{u_i(k)\} \\ z(k) &= \mathrm{col}_N\{z_i(t)\},\ \bar{z}(k) = \mathrm{col}_N\{\bar{z}_i(t)\}. \end{aligned}$$

Combining the consensus protocol (3.6) with the multi-agent systems (3.7), we have the following compact form:

$$\begin{aligned} x(k+1) = {} & (I_N \otimes A_k)x(k) + (I_N \otimes B_k K_k)e^t(k) \\ & + (I_N \otimes B_k K_k)\xi^*(k) + (I_N \otimes D_k)w(k). \end{aligned} \tag{3.11}$$

It can be verified that $\xi^*(k) = -(H \otimes I_{n_x})x(k)$. Letting $s(k) \triangleq e^t(k) - e^s(k) = \xi^t(k) - \xi(k)$ represent the control error introduced by the event-based schedule, (3.11) becomes

$$\begin{aligned} x(k+1) = {} & (I_N \otimes A_k - H \otimes B_k K_k)x(k) + (I_N \otimes B_k K_k)e^s(k) \\ & + (I_N \otimes B_k K_k)s(k) + (I_N \otimes D_k)w(k). \end{aligned}$$

Next, it follows from the relationship $e^s(k) = \xi(k) - \xi^*(k)$ that

$$\begin{aligned} e^s(k+1) = {} & (I_N \otimes (A_k - L_k C_k))\, e^s(k) \\ & - (H \otimes L_k E_k)v(k) + (H \otimes D_k)w(k). \end{aligned} \tag{3.12}$$

Similar to $\bar{z}_i(k)$, we let $\bar{x}_i(k) = x_i(k) - \frac{1}{N}\sum_{i=1}^{N} x_i(k)$ and then get $\bar{x}(k) = (H_m \otimes I_{n_x})x(k)$, where $\bar{x}(k) = \mathrm{col}_N\{\bar{x}_i(k)\}$ and $H_m = I_N - \frac{1}{N}\mathbf{1}\mathbf{1}^T$. It can be derived that

$$\begin{aligned} \bar{x}(k+1) = {} & (H_m \otimes A_k - H_m H \otimes B_k K_k)x(k) + (H_m \\ & \otimes B_k K_k)e^s(k) + (H_m \otimes B_k K_k)s(k) + (H_m \otimes D_k)w(k). \end{aligned} \tag{3.13}$$

Utilizing the properties of matrix H_m, we have $H_m H_m = H_m$ as well as $H_m H = H H_m = H$. Apparently, (3.13) can be converted into the form

$$\begin{aligned} \bar{x}(k+1) = {} & (H_m \otimes A_k - H \otimes B_k K_k)\bar{x}(k) + (H_m \otimes B_k K_k)e^s(k) \\ & + (H_m \otimes B_k K_k)s(k) + (H_m \otimes D_k)w(k) \end{aligned}$$

By defining the variables $X(k) \triangleq [\ \bar{x}^T(k) \ \ (e^s(k))^T\ ]^T$ and $\omega(k) \triangleq [\ w^T(k) \ \ v^T(k)\ ]^T$, we obtain the augmented system as follows:

$$\begin{cases} X(k+1) = \mathcal{A}_k X(k) + \mathcal{B}_k s(k) + \mathcal{D}_k \omega(k) \\ \bar{z}(k) = \mathcal{M}_k X(k) \\ \xi(k) = \mathcal{H} X(k), \end{cases} \tag{3.14}$$

where

$$\mathcal{A}_k = \begin{bmatrix} H_m \otimes A_k - H \otimes B_k K_k & H_m \otimes B_k K_k \\ 0 & I_N \otimes (A_k - L_k C_k) \end{bmatrix},$$
$$\mathcal{B}_k = \begin{bmatrix} H_m \otimes B_k K_k \\ 0 \end{bmatrix}, \quad \mathcal{D}_k = \begin{bmatrix} H_m \otimes D_k & 0 \\ H \otimes D_k & -H \otimes L_k E_k \end{bmatrix},$$
$$\mathcal{M}_k = \begin{bmatrix} I_N \otimes M_k & 0 \end{bmatrix}, \quad \mathcal{H} = \begin{bmatrix} -H \otimes I_{n_x} & I_N \otimes I_{n_x} \end{bmatrix}.$$

We are now in a position to state the problem addressed in this chapter as follows. We aim to design appropriate controller and estimator parameters to ensure that the controlled system (3.14) achieves the following H_∞ consensus performance constraint over the finite horizon $[0, T]$:

$$\|\bar{z}(k)\|^2_{[0,T]} < \gamma^2 \left\{ \|\omega(k)\|^2_{[0,T]} + \bar{x}^T(0)(I_N \otimes W)\bar{x}(0) \right\}. \tag{3.15}$$

3.2 Main Results

To start with, we first deal with the performance analysis problem, that is, derive the sufficient conditions under which the H_∞ consensus performance requirement (3.15) is guaranteed in terms of the feasibility of a backward RDE.

Lemma 3.1 *Consider the multi-agent systems* (3.1) *with the estimator-type consensus protocol* (3.5) *and the event-based mechanism* (3.10). *Given a disturbance attenuation level* $\gamma > 0$, *a positive scalar* $\theta > 0$, *and a positive definite matrix* W. *For any disturbance sequence* $\{\omega(k)\}_{0 \le k \le T}$, *the augmented system* (3.14) *satisfies the* H_∞ *consensus performance index if there exist a set of matrices* $\{K_k\}_{0 \le k \le T}$, $\{L_k\}_{0 \le k \le T}$ *and a set of nonnegative definite matrices* $\{P_k\}_{0 \le k \le T+1}$ *(with final condition* $P_{T+1} = 0$*) to the following backward RDE:*

$$\begin{aligned} P_k = {} & \mathcal{A}_k^T R_{k+1} \mathcal{A}_k + \mathcal{A}_k^T R_{k+1} \mathcal{D}_k \Omega_{k+1}^{-1} \mathcal{D}_k^T R_{k+1} \mathcal{A}_k \\ & + \mathcal{M}_k^T \mathcal{M}_k + \sigma \theta^2 \mathcal{H}^T \mathcal{H} \end{aligned} \tag{3.16}$$

subject to

$$\begin{aligned}\Phi_{k+1} &\triangleq \theta^2 I - \mathcal{B}_k^T P_{k+1}\mathcal{B}_k > 0,\\ \Omega_{k+1} &\triangleq \gamma^2 I - \mathcal{D}_k^T R_{k+1}\mathcal{D}_k > 0,\\ P_0 &< \gamma^2(I_{2N}\otimes W),\end{aligned} \tag{3.17}$$

where

$$R_{k+1} \triangleq P_{k+1}(I - \theta^{-2}\mathcal{B}_k\mathcal{B}_k^T P_{k+1})^{-1}. \tag{3.18}$$

Proof First, define a Lyapunov-like quadratic function $V_k(X(k)) = X^T(k)P_kX(k)$. For a set of nonnegative definite matrices $\{P_k\}_{0\le k\le T}$, the difference of $V_k(X(k))$ along the trajectory of (3.14) is calculated as follows:

$$\begin{aligned}Y_k^{(1)} &\triangleq V_{k+1}(X(k+1)) - V_k(X(k))\\ &= \|\mathcal{A}_kX(k) + \mathcal{B}_ks(k) + \mathcal{D}_k\omega(k)\|^2_{P_{k+1}} - \|X(k)\|^2_{P_k}\\ &= X^T(k)\mathcal{A}_k^T P_{k+1}\mathcal{A}_kX(k) + 2X^T(k)\mathcal{A}_k^T P_{k+1}\mathcal{B}_ks(k) + 2X^T(k)\mathcal{A}_k^T P_{k+1}\mathcal{D}_k\omega(k)\\ &+ 2s^T(k)\mathcal{B}_k^T P_{k+1}\mathcal{D}_k\omega(k) + s^T(k)\mathcal{B}_k^T P_{k+1}\mathcal{B}_ks(k) + \omega^T(k)\mathcal{D}_k^T P_{k+1}\mathcal{D}_k\omega(k) - X^T(k)P_kX(k).\end{aligned}$$

Furthermore, by introducing the zero term $\|\bar{z}(k)\|^2 - \gamma^2\|\omega(k)\|^2 - \|\bar{z}(k)\|^2 + \gamma^2\|\omega(k)\|^2 + \theta^2\|s(k)\|^2 - \theta^2\|s(k)\|^2$, we obtain that

$$\begin{aligned}Y_k^{(1)} &= X^T(k)(\mathcal{A}_k^T P_{k+1}\mathcal{A}_k + \mathcal{M}_k^T\mathcal{M}_k - P_k)X(k) - \|s(k)\|^2_{\Phi_{k+1}}\\ &- \|\omega(k)\|^2_{\gamma^2 I - \mathcal{D}_k^T P_{k+1}\mathcal{D}_k} + 2X^T(k)\mathcal{A}_k^T P_{k+1}\mathcal{D}_k\omega(k)\\ &+ 2s^T(k)\mathcal{B}_k^T P_{k+1}\mathcal{D}_k\omega(k) + 2X^T(k)\mathcal{A}_k^T P_{k+1}\mathcal{B}_ks(k)\\ &- \|\bar{z}(k)\|^2 + \gamma^2\|\omega(k)\|^2 + \theta^2\|s(k)\|^2.\end{aligned}$$

By using the matrix inversion lemma and substituting Φ_{k+1}, one has

$$R_{k+1} = P_{k+1} + P_{k+1}\mathcal{B}_k\Phi_{k+1}^{-1}\mathcal{B}_k^T P_{k+1} = P_{k+1}(I - \theta^{-2}\mathcal{B}_k\mathcal{B}_k^T P_{k+1})^{-1}.$$

Completing the square for $s(k)$, we have the following equation:

$$\begin{aligned}Y_k^{(1)} &= X^T(k)(\mathcal{A}_k^T R_{k+1}\mathcal{A}_k + \mathcal{M}_k^T\mathcal{M}_k - P_k)X(k)\\ &+ 2X^T(k)\mathcal{A}_k^T R_{k+1}\mathcal{D}_k\omega(k) - \|s(k) - s^*(k)\|^2_{\Phi_{k+1}}\\ &- \|\omega(k)\|^2_{\Omega_{k+1}} - \|\bar{z}(k)\|^2 + \gamma^2\|\omega(k)\|^2 + \theta^2\|s(k)\|^2,\end{aligned}$$

where $s^*(k) \triangleq \Phi_{k+1}^{-1}\mathcal{B}_k^T P_{k+1}\mathcal{D}_k\omega(k) + \Phi_{k+1}^{-1}\mathcal{B}_k^T P_{k+1}\mathcal{A}_kX(k)$. Furthermore, considering the definition of Ω_{k+1} in (3.17) and letting $\omega^*(k) \triangleq \Omega_{k+1}^{-1}\mathcal{D}_k^T R_{k+1}\mathcal{A}_kX(k)$, we complete the square for $\omega(k)$ as follows:

$$\begin{aligned}Y_k^{(1)} &= X^T(k)(\mathcal{A}_k^T R_{k+1}\mathcal{A}_k + \mathcal{M}_k^T\mathcal{M}_k + \mathcal{A}_k^T R_{k+1}^T\mathcal{D}_k\Omega_{k+1}^{-1}\\ &\times \mathcal{D}_k^T R_{k+1}\mathcal{A}_k - P_k)X(k) - \|s(k) - s^*(k)\|^2_{\Phi_{k+1}} - \|\bar{z}(k)\|^2\\ &+ \gamma^2\|\omega(k)\|^2 + \theta^2\|s(k)\|^2 - \|\omega(k) - \omega^*(k)\|^2_{\Omega_{k+1}}.\end{aligned}$$

Without loss of generality, the initial estimation error can be chosen as zero and let us consider a performance index

$$\begin{aligned} & J_1(K_k, L_k, \omega(k), s(k)) \\ & \triangleq \|\bar{z}(k)\|^2_{[0,T]} - \gamma^2\|\omega(k)\|^2_{[0,T]} - \gamma^2 X^T(0)(I_{2N} \otimes W)X(0). \end{aligned}$$

According to the triggering inequality (3.9), one has

$$s^T(k)s(k) \le \sigma\xi^T(k)\xi(k) = \sigma X^T(k)\mathcal{H}^T\mathcal{H}X(k).$$

Then, it follows from the conditions $\Phi_{k+1} > 0$, $\Omega_{k+1} > 0$, $P_0 < \gamma^2(I_{2N} \otimes W)$, and the final condition $P_{T+1} = 0$ that

$$\begin{aligned} & J_1(K_k, L_k, \omega(k), s(k)) \\ & \le \sum_{k=0}^{T} Y_k^{(1)} + \sum_{k=0}^{T} \left(\|\bar{z}(k)\|^2 - \gamma^2\|\omega(k)\|^2\right) \\ & \le \sum_{k=0}^{T} \Big\{X^T(k)(\mathcal{A}_k^T R_{k+1}\mathcal{A}_k + \mathcal{M}_k^T\mathcal{M}_k + \mathcal{A}_k^T R_{k+1}\mathcal{D}_k\Omega_{k+1}^{-1}\mathcal{D}_k^T R_{k+1}\mathcal{A}_k \\ & \quad + \sigma\theta^2\mathcal{H}^T\mathcal{H} - P_k)X(k)\Big\} - \sum_{k=0}^{T}\Big\{\|(s(k) - s^*(k)\|^2_{\Phi_{k+1}} + \|\omega(k) - \omega^*(k)\|^2_{\Omega_{k+1}}\Big\} \\ & \le \sum_{k=0}^{T}\Big\{ - \|(s(k) - s^*(k)\|^2_{\Phi_{k+1}} - \|\omega(k) - \omega^*(k)\|^2_{\Omega_{k+1}}\Big\} < 0, \end{aligned}$$

where (3.16) and (3.17) have been used in deriving the last inequality (3.19). To this end, it can be concluded that the H_∞ consensus of multi-agent systems is achieved.

Since the feasibility of (3.16) subject to (3.17) is difficult to tackle directly, let us now propose an approach for computing the controller parameters K_k and estimator parameters L_k in each step under the worst situation, i.e., $\omega(k) = \omega^*(k) = \Omega_{k+1}^{-1}\mathcal{D}_k^T R_{k+1}\mathcal{A}_k X(k)$ and $s(k) = s^*(k) = \Delta_{k+1}X(k)$ with

$$\Delta_{k+1} \triangleq \Phi_{k+1}^{-1}\mathcal{B}_k^T(P_{k+1}\mathcal{D}_k\Omega_{k+1}^{-1}\mathcal{D}_k^T R_{k+1} + P_{k+1})\mathcal{A}_k.$$

In the sequel, we rewrite the augmented system (3.14) as follows:

$$X(k+1) = (\bar{\mathcal{A}}_k + \mathcal{D}_k\Omega_{k+1}^{-1}\mathcal{D}_k^T R_{k+1}\mathcal{A}_k)X(k) + \bar{\mathcal{B}}_k u^t(k) \tag{3.19}$$

where $\bar{\mathcal{A}}_k \triangleq \text{diag}\{H_m \otimes A_k, I_N \otimes (A_k - L_k C_k)\}$ and $\bar{\mathcal{B}}_k \triangleq [\ H_m \otimes B_k^T \ \ 0\ ]^T$.

Before the statement of Lemma 2, we introduce the following notations in order to simplify the presentation:

$$\tilde{\bar{\mathcal{A}}}_k \triangleq \text{diag}\{H_m \otimes A_k, I_N \otimes A_k\}, \mathcal{K}_k \triangleq I_N \otimes K_k, \mathcal{L}_k \triangleq I_N \otimes L_k,$$
$$\bar{\bar{\mathcal{A}}}_k \triangleq \text{diag}\{\mathbf{0}_N \otimes I_{n_x}, -I_N \otimes L_k C_k\}, \ \mathbf{I} = [\,0 \ \ I_{n_x N}\,]^T$$

Lemma 3.2 *Consider the multi-agent systems* (3.1) *with the estimator-type consensus protocol* (3.5), *and the event-based mechanism* (3.10). *Let the disturbance attenuation level* $\gamma > 0$, *positive scalars* $\theta > 0$, $\varepsilon_1 > 0$, *and the positive definite matrix* W *be given. For the worst disturbance sequence* $\{\omega^*(k)\}_{0 \le k \le T}$ *and control error sequence* $\{s^*(k)\}_{0 \le k \le T}$, *the augmented system* (3.14) *satisfies the* H_∞ *consensus performance requirement if there exist solutions* $(P_k, \mathcal{Q}_k, \mathcal{K}_k, \mathcal{L}_k)$ *to the following backward RDEs:*

$$\begin{cases} P_k = \mathcal{A}_k^T R_{k+1} \mathcal{A}_k + \mathcal{A}_k^T R_{k+1}^T \mathcal{B}_k \Omega_{k+1}^{-1} \mathcal{B}_k^T R_{k+1} \mathcal{A}_k \\ \quad +\mathcal{M}_k^T \mathcal{M}_k + \sigma\theta^2 \mathcal{H}^T \mathcal{H} \\ P_k \ge 0, \quad P_{T+1} = 0 \end{cases} \tag{3.20}$$

and

$$\begin{cases} \mathcal{Q}_k = (\bar{\mathcal{A}}_k + \mathcal{D}_k \Omega_{k+1}^{-1} \mathcal{D}_k^T R_{k+1} \mathcal{A}_k)^T \mathcal{Q}_{k+1} (\bar{\mathcal{A}}_k + \mathcal{D}_k \Omega_{k+1}^{-1} \\ \quad \times \mathcal{D}_k^T R_{k+1} \mathcal{A}_k) + \mathcal{M}_k^T \mathcal{M}_k + \Delta_{k+1}^T \mathcal{K}_k^T \Pi_{k+1} \mathcal{K}_k \Delta_{k+1} \\ \quad - \tilde{\bar{\mathcal{A}}}_k^T \mathcal{Q}_{k+1} \bar{\mathcal{B}}_k \Pi_{k+1}^{-1} \bar{\mathcal{B}}_k^T \mathcal{Q}_{k+1}^T \tilde{\bar{\mathcal{A}}}_k + \Xi_{k+1}^{(1)} + \Xi_{k+1}^{(1)T} \\ \quad - \varepsilon_1^{-1} \boldsymbol{I} \Xi_{k+1}^{(2)T} \Xi_{k+1}^{(2)} \boldsymbol{I}^T \\ \mathcal{Q}_{T+1} = 0 \end{cases} \tag{3.21}$$

subject to

$$\begin{cases} \Phi_{k+1} \triangleq \theta^2 I - \mathcal{B}_k^T P_{k+1} \mathcal{B}_k > 0 \\ \Omega_{k+1} \triangleq \gamma^2 I - \mathcal{D}_k^T R_{k+1} \mathcal{D}_k > 0 \\ \Pi_{k+1} \triangleq \bar{\mathcal{B}}_k^T \mathcal{Q}_{k+1} \bar{\mathcal{B}}_k + I > 0 \\ P_0 < \gamma^2 (I_{2N} \otimes W). \end{cases} \tag{3.22}$$

$\Xi_{k+1}^{(1)}$ *and* $\Xi_{k+1}^{(2)}$ *are defined as follows:*

$$\begin{aligned} \Xi_{k+1}^{(1)} &\triangleq (\mathcal{D}_k \Omega_{k+1}^{-1} \mathcal{D}_k^T R_{k+1} \mathcal{A}_k)^T \mathcal{Q}_{k+1} \bar{\mathcal{B}}_k \mathcal{K}_k \mathcal{H} + (\bar{\mathcal{A}}_k + \mathcal{D}_k \Omega_{k+1}^{-1} \\ &\quad \times \mathcal{D}_k^T P_{k+1} \mathcal{A}_k)^T \mathcal{Q}_{k+1} \bar{\mathcal{B}}_k \mathcal{K}_k \Delta_{k+1} + \mathcal{H}^T \mathcal{K}_k^T \Pi_{k+1} \mathcal{K}_k \Delta_{k+1} \\ &\quad + \boldsymbol{I} (I_N \otimes L_k C_k)^T \Xi_{k+1}^{(2)} (H \otimes I_{n_x}) [I_{n_x N} \ \ 0] \\ \Xi_{k+1}^{(2)} &\triangleq \mathcal{Q}_{k+1}^{(21)} (H_m \otimes B_k K_k), \end{aligned}$$

where $Q_{k+1}^{(ij)}(i, j = 1, 2)$ *is the block elements of the matrix* Q_{k+1} *with appropriate dimensions.*

Proof Define a cost functional as $J_2(K_k, L_k, \omega^*(k), s^*(k)) \triangleq \|\bar{z}(k)\|^2_{[0,T]} + \|u^t(k)\|^2_{[0,T]} + \varepsilon_1\|\tilde{e}^s(k)\|^2_{[0,T]}$, where $\tilde{e}^s(k) \triangleq (I_N \otimes L_k C_k)e^s(k)$, and ε_1 is introduced for more flexibility in the estimator parameter design. Furthermore, introducing the function $Y_k^{(2)} \triangleq X^T(k+1)Q_{k+1}X(k+1) - X^T(k)Q_k X(k)$, it follows from (3.19) that

$$\begin{aligned}
Y_k^{(2)} = {} & X^T(k)(\bar{\mathcal{A}}_k + \mathcal{D}_k\Omega_{k+1}^{-1}\mathcal{D}_k^T R_{k+1}\mathcal{A}_k)^T Q_{k+1}(\bar{\mathcal{A}}_k + \mathcal{D}_k\Omega_{k+1}^{-1}\mathcal{D}_k^T R_{k+1}\mathcal{A}_k)X(k) \\
& - X^T(k)Q_k X(k) + 2X^T(k)(\bar{\mathcal{A}}_k + \mathcal{D}_k\Omega_{k+1}^{-1}\mathcal{D}_k^T R_{k+1}\mathcal{A}_k)^T Q_{k+1}\bar{\mathcal{B}}_k u^t(k) \\
& + (u^t(k))^T \bar{\mathcal{B}}_k^T Q_{k+1}\bar{\mathcal{B}}_k u^t(k)
\end{aligned}$$

which leads to

$$\begin{aligned}
& J_2(K_k, L_k, \omega^*(k), s^*(k)) \\
& = \sum_{k=0}^{T}\left(Y_k + \|\bar{z}(k)\|^2 + \|u^t(k)\|^2 + \varepsilon_1\|\tilde{e}^s(k)\|^2\right) + X^T(0)Q_0X(0) \\
& = \sum_{k=0}^{T}\Big\{X^T(k)\Big[(\bar{\mathcal{A}}_k + \mathcal{D}_k\Omega_{k+1}^{-1}\mathcal{D}_k^T R_{k+1}\mathcal{A}_k)^T Q_{k+1}(\bar{\mathcal{A}}_k + \mathcal{D}_k\Omega_{k+1}^{-1}\mathcal{D}_k^T R_{k+1}\mathcal{A}_k) \\
& \quad + \mathcal{M}_k^T\mathcal{M}_k - Q_k\Big]X(k) + 2X^T(k)(\bar{\mathcal{A}}_k + \mathcal{D}_k\Omega_{k+1}^{-1}\mathcal{D}_k^T R_{k+1}\mathcal{A}_k)^T Q_{k+1}\bar{\mathcal{B}}_k u^t(k) \\
& \quad + (u^t(k))^T(\bar{\mathcal{B}}_k^T Q_{k+1}\bar{\mathcal{B}}_k + I)u^t(k) + \varepsilon_1\tilde{e}^{sT}(k)\tilde{e}^s(k)\Big\} + X^T(0)Q_0X(0).
\end{aligned}$$

Noting the fact that $u^t(k) = u(k) + \mathcal{K}_k s(k)$, one has

$$\begin{aligned}
& J_2(K_k, L_k, \omega^*(k), s^*(k)) \\
& = \sum_{k=0}^{T}\Big\{X^T(k)\Big[(\bar{\mathcal{A}}_k + \mathcal{D}_k\Omega_{k+1}^{-1}\mathcal{D}_k^T R_{k+1}\mathcal{A}_k)^T Q_{k+1}(\bar{\mathcal{A}}_k + \mathcal{D}_k\Omega_{k+1}^{-1}\mathcal{D}_k^T R_{k+1}\mathcal{A}_k) \\
& + \mathcal{M}_k^T\mathcal{M}_k - Q_k\Big]X(k) + 2X^T(k)(\bar{\mathcal{A}}_k + \mathcal{D}_k\Omega_{k+1}^{-1}\mathcal{D}_k^T R_{k+1}\mathcal{A}_k)^T Q_{k+1} \\
& + X^T(k)\Delta_{k+1}^T\mathcal{K}_k^T\Pi_{k+1}\mathcal{K}_k\Delta_{k+1}X(k) + 2u^T(k)\Pi_{k+1}\mathcal{K}_k\Delta_{k+1}X(k) \\
& + 2X^T(k)(\mathcal{D}_k\Omega_{k+1}^{-1}\mathcal{D}_k^T R_{k+1}\mathcal{A}_k)^T Q_{k+1}\bar{\mathcal{B}}_k u(k) + 2X^T(k)(\bar{\mathcal{A}}_k + \tilde{\bar{\mathcal{A}}}_k)^T Q_{k+1}\bar{\mathcal{B}}_k u(k) \\
& \times \bar{\mathcal{B}}_k\mathcal{K}_k\Delta_{k+1}X(k)\Big\} + X^T(0)Q_0X(0) + \varepsilon_1\|\tilde{e}^s(k)\|^2 + \|u(k)\|^2_{\Pi_{k+1}}.
\end{aligned}$$

Completing the square with respect to $\tilde{e}^s(k)$ and $u(k)$, it follows from (3.21) that

$$
\begin{aligned}
&J_2(K_k, L_k, \omega^*(k), s^*(k)) \\
&= \sum_{k=0}^{T} \left\{ \varepsilon_1 \| \tilde{e}^s(k) - \varepsilon_1^{-1} \Xi_{k+1}^{(2)} e^s(k) \|^2 + \| u(k) + \Pi_{k+1}^{-1} \bar{\mathcal{B}}_k^T Q_{k+1}^T \tilde{\bar{\mathcal{A}}}_k X(k) \|_{\Pi_{k+1}}^2 \right\} \\
&\quad + X^T(0) Q_0 X(0),
\end{aligned} \tag{3.23}
$$

which ends the proof.

Remark 3.4 Under the constraint of the event rules, the error $s(k)$ may not always be taken as the worst case $s^*(k)$ during the dynamics evolution. In other words, some non-worst cases may happen during the process. Fortunately, it is clear from (3.19) that, with the solutions $(P_k, Q_k, \mathcal{K}_k, \mathcal{L}_k)$ to (3.20) and (3.21) subject to (3.22), the H_∞ consensus performance requirement (3.15) can be satisfied even in the non-worst cases.

In the following theorem, an explicit algorithm is given to compute the controller parameters K_k and estimator parameters L_k in each step of the time-varying consensus process.

Theorem 3.2.1 *Consider the multi-agent systems* (3.1) *with the estimator-type consensus protocol* (3.5) *and the event-based mechanism* (3.10). *Let the disturbance attenuation level* $\gamma > 0$, *positive scalars* $\theta > 0$, $\varepsilon_1 > 0$, *and the positive definite matrix* W *be given. The augmented system* (3.14) *satisfies the* H_∞ *consensus performance requirement if there exist solutions* (P_k, Q_k, K_k^*, L_k^*) *to the backward RDEs* (3.20) *and* (3.21) *subject to* (3.22) *with the controller and estimator parameters given as follows:*

$$
\begin{aligned}
K_k^* &= -[\delta_k^{(1)}, \delta_k^{(2)}, \cdots, \delta_k^{(N)}][h^{(1)}, h^{(2)}, \cdots, h^{(N)}]^\dagger \\
L_k^* &= [\,\kappa_k^{(1)}, \kappa_k^{(2)}, \cdots, \kappa_k^{(N)}\,][\,{}_k^{(1)}, {}_k^{(2)}, \cdots, {}_k^{(N)}\,]^\dagger,
\end{aligned} \tag{3.24}
$$

where

$$
\begin{cases}
\mathscr{D}_k \triangleq \Pi_{k+1}^{-1} \bar{\mathcal{B}}_k^T Q_{k+1} \tilde{\bar{\mathcal{A}}}_k \triangleq [\delta_k^{(1)T}, \delta_k^{(2)T}, \cdots, \delta_k^{(N)T}]^T \\
\mathcal{H} \triangleq [h^{(1)T}, h^{(2)T}, \cdots, h^{(N)T}]^T \\
\mathscr{V}_k \triangleq (I_N \otimes C_k) \triangleq [\,{}_k^{(1)T}, {}_k^{(2)T}, \cdots, {}_k^{(N)T}\,]^T \\
\mathscr{W}_k \triangleq \varepsilon_1^{-1} Q_{k+1}^{(21)} (H_m \otimes B_k K_k) \triangleq [\,\kappa_k^{(1)T}, \kappa_k^{(2)T}, \cdots, \kappa_k^{(N)T}\,]^T.
\end{cases}
$$

Proof It is easy to verify that the best choice of the controller parameter $\mathcal{K}_k = I_N \otimes K_k$ and the estimator parameter $\mathcal{L}_k = I_N \otimes L_k$ that suppress the cost function (3.23) is determined in each iteration backward as follows:

$$
\begin{aligned}
\mathcal{K}_k^* &= \arg\min_{\mathcal{K}_k} \ \text{norm}\ (\mathcal{K}_k \mathcal{H} + \Pi_{k+1}^{-1} \bar{\mathcal{B}}_k^T Q_{k+1} \tilde{\bar{\mathcal{A}}}_k), \\
\mathcal{L}_k^* &= \arg\min_{L_k} \ \text{norm}\ (\mathcal{L}_k (I_N \otimes C_k) - \varepsilon_1^{-1} Q_{k+1}^{(21)} (H_m \otimes B_k K_k)).
\end{aligned}
$$

The controller parameter can be rearranged as

Table 3.1 The H_∞ consensus control algorithm

Algorithm:	
Step 1	Set $k = T$, then $P_{T+1} = Q_{T+1} = 0$ are available
Step 2	Calculate the matrices Φ_{k+1}, Ω_{k+1}, and Π_{k+1} by (3.22), respectively
Step 3	If Φ_{k+1} and Ω_{k+1} are all positive definite, then we can obtain the controller gain K_k and estimator gain L_k by (3.24) and step to the next procedure, else jump to *Step 6*
Step 4	Solve the backward RDEs of (3.20) and (3.21) to get P_k and Q_k
Step 5	If $k \neq 0$, set $k = k - 1$ and go back to *Step 2*, else turn to the next step
Step 6	If $\Phi_{k+1} \leq 0$, or $\Omega_{k+1} \leq 0$, or $\Pi_{k+1} \leq 0$ or $P_0 \geq \gamma^2 W$, this algorithm is infeasible. Stop

$$K_k^* = \arg \min_{K_k} \text{ norm } (K_k[h^{(1)}, h^{(2)}, \cdots, h^{(N)}] + [\delta_k^{(1)}, \delta_k^{(2)}, \cdots, \delta_k^{(N)}]),$$

and therefore

$$K_k^* = -[\delta_k^{(1)}, \delta_k^{(2)}, \cdots, \delta_k^{(2N)}][h^{(1)}, h^{(2)}, \cdots, h^{(N)}]^\dagger.$$

Similarly, we have

$$L_k^* = \arg \min_{L_k} \text{ norm } (L_k[\ {}_k^{(1)}, {}_k^{(2)}, \cdots, {}_k^{(N)}\ ] - [\ \kappa_k^{(1)}, \kappa_k^{(2)}, \cdots, \kappa_k^{(N)}\ ]).$$

By using the Moore–Penrose pseudo inverse, we can easily determine the estimator parameter as $L_k^* = [\kappa_k^{(1)}, \kappa_k^{(2)}, \cdots, \kappa_k^{(N)}][{}_k^{(1)}, {}_k^{(2)}, \cdots, {}_k^{(N)}]^\dagger$. With the designed K_k^* and L_k^*, it follows from (3.20) and (3.22) that the performance index $J_1 < 0$. Therefore, the multi-agent system achieves the H_∞ consensus performance constraint over the finite horizon $[0, T]$. The proof is now complete.

In Theorem 3.2.1, a backward RDE criterion is established to guarantee the H_∞ consensus performance constraint. By recursively solving the other backward RDE and resorting to the Moore–Penrose pseudo inverse, the explicit expressions of controller and estimator parameters can be obtained. Finally, let us outline the algorithm in Table 3.1.

Remark 3.5 In this chapter, we investigate finite-horizon behaviors of multi-agent systems under the event-based control. The horizon length T, which should be selected to meet the transient performance requirement of the systems.

3.3 An Illustrative Example

To illustrate the effectiveness of the obtained theoretical results, we apply the event-based consensus control to an example system by numerical simulation in this section.

As shown in Fig. 3.1, the topology of the multi-agent system can be represented by undirected graph $\mathcal{G} = (\mathcal{V}, \mathcal{E}, \mathcal{A})$ with the set of the nodes $\mathcal{V} = \{1, 2, 3, 4\}$, set of edges

$$\mathcal{E} = \{(1, 2), (1, 3), (2, 1), (2, 3), (3, 1), (3, 2), (3, 4), (4, 3)\}.$$

The individual dynamics of agents are given by (3.7) with the following parameters:

$$A_k = \begin{bmatrix} 0.98 + 0.04\sin(0.12k) & 0.4 \\ 0.15 & -0.75 + 0.2\cos(0.1k) \end{bmatrix}$$

$$\begin{aligned} B_k &= [\, 0.8 + 0.2\sin(0.4k) \;\; 0.5 \,]^T \quad E_k = 0.04 \\ C_k &= [\, 0.82 \;\; 0.62 + 0.35\cos(0.3k) \,] \quad M_k = [\, 0.7 \; -0.64 \,] \\ D_k &= [0.16 + 0.05\cos(0.3k) \;\; 0.18 \,]^T. \end{aligned}$$

The process and measurement disturbances belonging to $\mathcal{L}_2[0, \; T]$ are selected as random variables uniformly distributed in the region $[-0.05, 0.05]$ and $[-0.5, 0.5]$, respectively. In this simulation, we choose the thresholds $\sigma = 0.28$, and the scalars $\varepsilon_1 = 0.01$ and $\theta = 1.8$. The H_∞ performance index γ, the positive definite matrix W, and the time horizon T are taken as 5, $\text{diag}_2\{2.8, 2.8\}$, 80, respectively. The initial positions of four agents are uniformly distributed between -5 and 5. According to Theorem 3.2.1, the H_∞ performance index for the multi-agent system can be

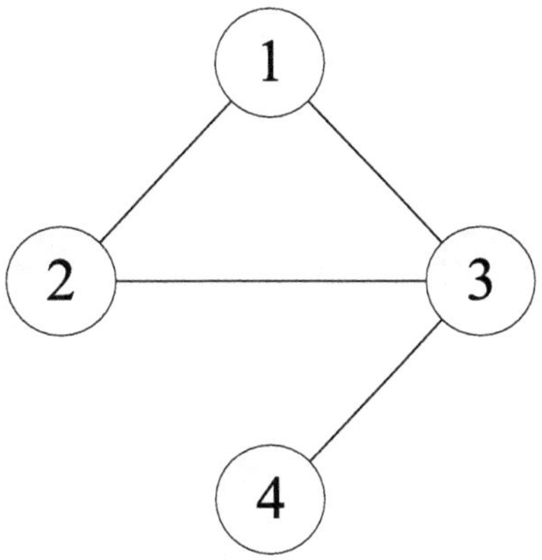

Fig. 3.1 The graph

Table 3.2 The controller and estimator gain matrices

Time k	K_k	L_k^T
1	[−0.13130.0169]	[0.1785 0.0332]
2	[−0.13000.0172]	[0.1666 0.0319]
3	[−0.12870.0176]	[0.1559 0.0307]
4	[−0.12730.0180]	[0.1464 0.0299]
...	...	...

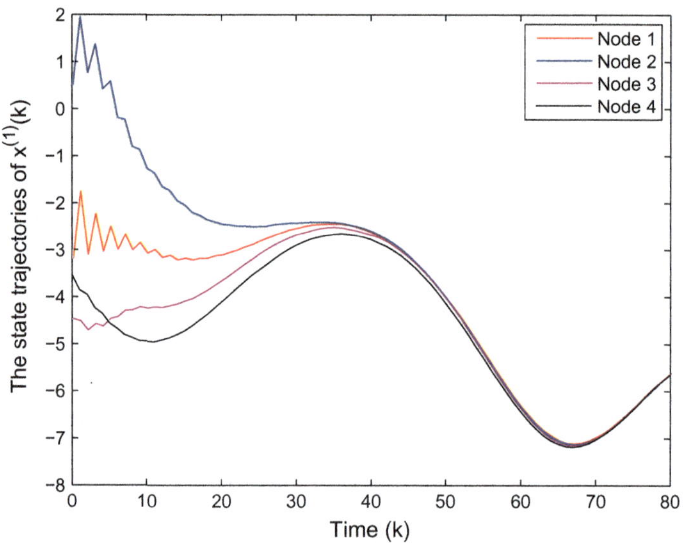

Fig. 3.2 The state trajectories $x_i^{(1)}(k)$

guaranteed with the controller parameters K_k and estimation parameters L_k computed in each iteration, see Table 3.2. Simulation results are presented in Figs. 3.2, 3.3, 3.4, and 3.5. Figures 3.2 and 3.3 depict the state trajectories of $x_i(k)$ $(i = 1, 2, 3, 4)$ whose j-th element is denoted by $x_i^{(j)}(k)$ $(j = 1, 2)$. The consensus error $\bar{z}(k)$ can be found in Fig. 3.4 from which we can see that the time-varying multi-agent system achieves H_∞ consensus over a finite horizon. In addition, Fig. 3.5 shows that the execution frequencies (for information broadcast and actuator adjustments) are dramatically decreased. As a result, the superiority of the proposed event-based mechanism is clearly shown.

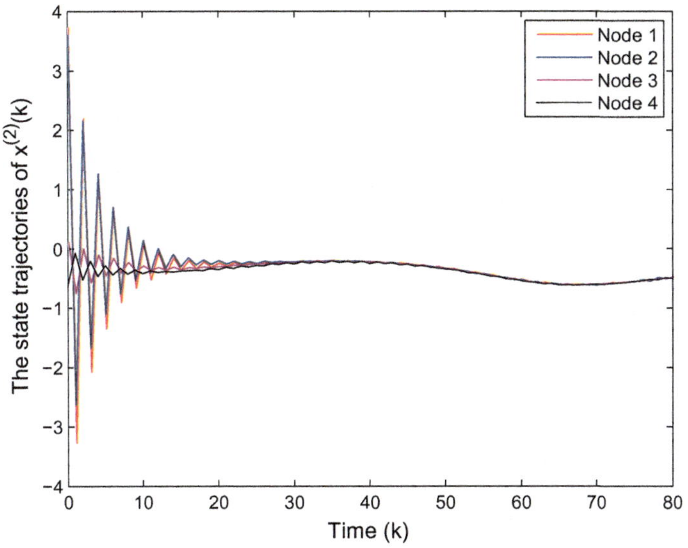

Fig. 3.3 The state trajectories $x_i^{(2)}(k)$

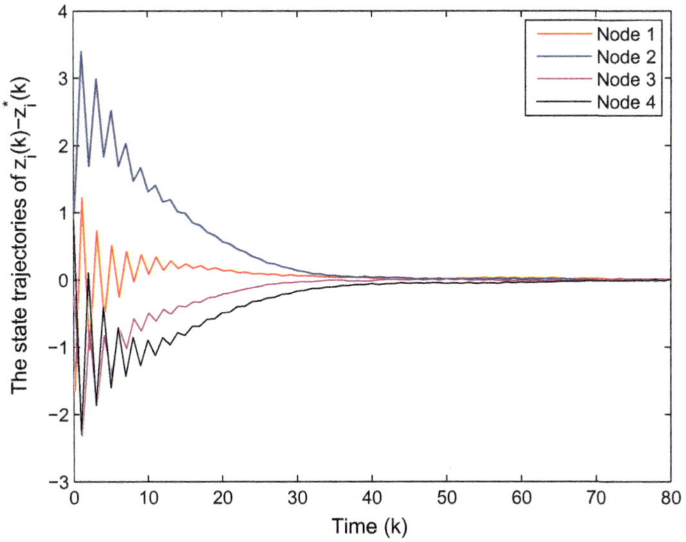

Fig. 3.4 The consensus error $\bar{z}_i(k)$

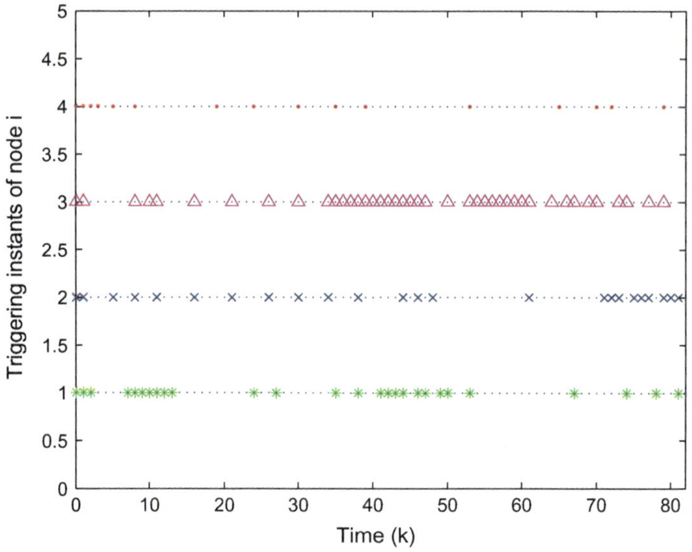

Fig. 3.5 The triggering instants

3.4 Conclusions

This chapter has addressed the consensus problem for discrete time-varying multi-agent systems with external disturbances. The definition of H_∞ consensus has been introduced to characterize the transient performance of the coordination. An event-based estimator-based output feedback protocol has been proposed to generate the control signals. Subsequently, by utilizing the H_∞ analysis techniques, a set of RDEs has been derived to determine whether the H_∞ performance constraint is met, and then the appropriate H_∞ controller as well as the estimator parameters have been designed under the worst situations. In the end, an illustrative example has been presented to demonstrate the effectiveness of the theoretical results proposed in this chapter.

References

1. J.R. Carpenter, Decentralized control of satellite formations. Int. J. Robust Nonlinear Control **12**(2/3), 141–161 (2002)
2. H. Su, X. Wang, Z. Lin, Flocking of multi-agents with a virtual leader. IEEE Trans. Autom. Control **54**(2), 293–307 (2009)

3. J.R. Lawton, R.W. Beard, Synchronized multiple spacecraft rotations. Automatica **38**(8), 1359–1364 (2002)
4. M.J. Garcia-Ligero, A. Hermoso-Carazo, J. Linares-Perez, Distributed and centralized fusion estimation from multiple sensors with Markovian delays. Appl. Math. Comput. **219**(6), 2932–2948 (2012)
5. J.G. Bender, An overview of systems studies of automated highway systems. IEEE Trans. Vehicular Technol. **40**(1), 82–99 (1991)
6. W. Ren, E. Atkins, Distributed multi-vehicle coordinated control via local information exchange. Int. J. Robust Nonlinear Control **17**(10–11), 1002–1033 (2007)
7. S.S. Stankovic, M.S. Stankovic, D.M. Stipanovic, Decentralized parameter estimation by consensus based stochastic approximation. IEEE Trans. Autom. Control **56**(3), 531–543 (2011)
8. L. Ding, Q.-L. Han, G. Guo, Network-based leader-following consensus for distributed multi-agent systems. Automatica **49**(7), 2281–2286 (2013)
9. L. Li, D.W.C. Ho, J. Lu, A unified approach to practical consensus with quantized data and time delay. IEEE Trans. Circ. Syst. Part I **60**(10), 2668–2678 (2013)
10. Y. Hu, J. Lam, J. Liang, Consensus control of multi-agent systems with missing data in actuators and Markovian communication failure. Int. Syst. Sci. **44**(10), 1867–1878 (2013)
11. K. You, Z. Li, L. Xie, Consensus condition for linear multi-agent systems over randomly switching topologies. Automatica **49**(10), 3125–3132 (2013)
12. Z. Li, W. Ren, X. Liu, L. Xie, Distributed consensus of linear multi-agent systems with adaptive dynamic protocols. Automatica **49**(7), 1986–1995 (2013)
13. J. Ghommam, M.S. Mahmoud, M. Saad, Robust cooperative control for a group of mobile robots with quantized information exchange. J. Frankl. Inst. **350**(8), 2291–2321 (2013)
14. N. Ilic, S.S. Stankovic, M.S. Stankovic, K.H. Johansson, consensus based distributed change detection using generalized likelihood ratio methodology. Signal Process. **92**(7), 1715–1728 (2012)
15. Q. Liu, Z. Wang, X. He, D.H. Zhou, A survey of event-based strategies on control and estimation. Syst. Sci. Control Eng. **2**(1), 90–97 (2014)
16. B. Shen, Z. Wang, Y.S. Hung, Distributed H_∞-consensus filtering in sensor networks with multiple missing measurements: the finite-horizon case. Automatica **46**(10), 1682–1688 (2010)
17. Z. Wang, H. Dong, B. Shen, H. Gao, Finite-horizon H_∞ filtering with missing measurements and quantization effects. IEEE Trans. Autom. Control **58**(7), 1707–1718 (2013)
18. K.J. Åström, B.M. Bernhardsson, Comparison of Riemann and Lebesgue sampling for first order stochastic systems, in *Proceedings of 41st IEEE Conference Decision and Control*, Las Vegas, NV, USA, vol. 2, pp. 2011–2016, December 2002
19. M. Lemmon, Event-triggered feedback in control, estimation, and optimization, in *Networked Control Systems* (Springer, London, 2010), pp. 293–358
20. X. Wang, M.D. Lemmon, Event-triggering in distributed networked control systems. IEEE Trans. Autom. Control **56**(3), 586–601 (2011)
21. J. Hu, G. Chen, H.X. Li, Distributed event-triggered tracking control of leader-follower multi-agent systems with communication delays. Kybernetika **47**(4), 630–643 (2011)
22. D.V. Dimarogonas, E. Frazzoli, K.H. Johansson, Distributed event-triggered control for multi-agent systems. IEEE Trans. Autom. Control **57**(5), 1291–1297 (2012)
23. X. Yin, D. Yue, S. Hu, Distributed event-triggered control of discrete-time heterogeneous multi-agent systems. J. Frankl. Inst. **350**(3), 651–669 (2013)
24. G.S. Seyboth, D.V. Dimarogonas, K.H. Johansson, Event-based broadcasting for multi-agent average consensus. Automatica **49**(1), 245–252 (2013)
25. Y. Fan, G. Feng, Y. Wang, C. Song, Distributed event-triggered control of multi-agent systems with combinational measurements. Automatica **49**(2), 671–675 (2012)
26. W. Zhu, Z. Jiang, G. Feng, Event-based consensus of multi-agent systems with general linear models. Automatica **50**(2), 552–558 (2014)
27. H. Yu, P.J. Antsaklis, Event-triggered output feedback control for networked control systems using passivity: Achieving L_2 stability in the presence of communication delays and signal quantization. Automatica **49**(1), 30–38 (2013)

28. H. Yu, P.J. Antsaklis, Output synchronization of networked passive systems with event-driven communication. IEEE Trans. Autom. Control **59**(3), 750–756 (2014)
29. P. Lin, Y. Jia and L. Li, Distributed robust H_∞ consensus control in directed networks of agents with time-delay. Syst. Control Lett. **57**(8), 643–653 (2008)
30. Y. Liu, Y. Jia, H_∞ consensus control of multi-agent systems with switching topology: a dynamic output feedback protocol. Int. J. Control **83**(3), 527–537 (2009)

Chapter 4
Event-Triggered Resilient Filtering with Measurement Quantization

The state estimation or filtering problem has long been a fundamental research issue in control engineering and signal processing with tremendous application insights in almost all practical systems especially in guidance, navigation, and vehicle control [1, 2]. Up to now, considerable research interest has been devoted to the networked state estimation problems, where the raw measurements are transmitted from sensors to a central unit for processing through a wired/wireless communication channel, see, for instance, [3].

The emergence and popularity of network communications have been posing new challenges for engineers and scientists when designing optimal state estimators. For most practical applications in networked environments, the network bandwidth for communication is exhaustible. Unnecessarily transmissions of the measurements from the sensor to the estimator might result in some undesirable phenomena and thus deteriorate the communication quality, which would reduce estimation accuracy. To handle such a resource limit issue, event-triggered communication, has recently drawn much research attention with the objective of reducing the number of network transmissions while maintaining certain system performance (e.g., stability for a control problem), see, for instance, [4–12] and the references therein. Specifically, in [10], the authors have presented the minimum bit-rates required to realize the event-triggered control system that is resilient to transient unexpected disturbances. As for event-triggered state estimation problems, in [12], the authors have proposed a variance-based event-triggering state estimation strategy whose idea is to decide whether an event should be triggered by calculating the error variance.

A critical assumption behind the conventional filter algorithm is that the filter has access to the accurate measurements of the plant. This assumption is, however, not true in certain situations. Note that sensors may sometimes experience abnormal working conditions or even failures, which give rise to inaccurate measurements suffering from random degradations. The research on filtering problems with degraded/missing measurements dates back as early as in [13], where the opti-

Q. Liu et al., *Stochastic Control and Filtering over Constrained Communication Networks*, Studies in Systems, Decision and Control 178,
https://doi.org/10.1007/978-3-030-00157-5_4

mal recursive filter has been designed for systems with missing measurements. On the other hand, for state estimation problem over communication networks, the raw measurements are usually quantized first before being transmitted. The gap between the quantized measurement and the raw one makes the conventional filter algorithm no longer applicable. A great number of results have been available in the recent literature on the topic of the networked control/estimation problems with discrete quantized measurements, see, e.g., [14–16].

It should be noted that another problem that we might encounter is the fluctuations or inaccuracies in the realization of filter/controller algorithms. According to [17], such inaccuracies arise from a variety of factors including, but are not limited to, finite-resolution measuring instruments, finite word length, round-off errors in numerical computation, and safe-tuning margin requirements. If not properly handled, the filter/controller gain variations might seriously deteriorate the system performance. As a result, particular research efforts have been paid on the design of *resilient* filters/controllers that are insensitive to the aforementioned inaccuracies [18, 19].

Motivated by the above discussion, the aim of this chapter is to develop event-triggered resilient filters for a class of networked systems subject to both quantization effects and random sensor failures. Generally speaking, the primary technical challenges we are encountering can be highlighted as threefold: *(1) due to the existence of the errors induced by quantization and event-triggered communication, the analytical expression of the error covariance is intractable and, as opposed to the traditional Kalman filtering case, there is no basis now for the subsequent minimum variance estimation; (2) although it is acknowledged that the increase of both the measurement inaccuracies and the event-triggering thresholds would inevitably reduce the filtering accuracy, there is a lack of rigorous mathematical proof as to the monotonicity which turns out to be a difficult task; and (3) the establishment of the convergence conditions for the designed steady-state optimal filter is a nontrivial objective as well.* In this chapter, we endeavor to tackle the identified three challenges. Particularly, a recursive Riccati-like difference equation is first constructed whose solution is a tight upper bound of the actual covariance, and the filter gain is then designed to minimize the upper bound as an alternative. By matrix transformations and inductive methods, we demonstrate that the minimized upper bound is a monotonically increasing function with respect to the thresholds. Furthermore, a sufficient condition for the convergence of the steady-state filter is established in terms of a matrix inequality.

4.1 Problem Formulation

Consider the following discrete-time system:

$$x(k+1) = Ax(k) + \alpha(k)h(x(k)) + w(k), \tag{4.1}$$

where $x(k) \in \mathbb{R}^n$ is the state vector of the plant, the process noise $w(k) \in \mathbb{R}^n$ is a Gaussian random vector with zero mean and covariance $Q > 0$, and $\alpha(k)$ is a scalar-valued random variable with zero mean and variance $\breve{\alpha}$. $h(x(k))$ is a nonlinear function to be defined later. The initial state $x(0)$ is a zero-mean Gaussian random vector with covariance $X_0 > 0$. The measurement equation subject to random sensor failures is given by

$$y_i(k) = \lambda_i(k)C_i x(k) + v_i(k), \ i = 1, 2, \ldots, m, \tag{4.2}$$

where $y_i(k) \in \mathbb{R}$ is the observation of ith sensor and the measurement noise $v_i(k) \in \mathbb{R}$ is a Gaussian random variable with zero mean and variance $R_i > 0$. $\lambda_i(k) \in \mathbb{R}$ is a random variable characterizing the random sensor failure, which takes values on the interval [0, 1] according to certain probabilistic density functions with mean $\bar{\lambda}_i$ and variance $\breve{\lambda}_i$. Additionally, A and C_i are known matrices of appropriate dimensions.

Define

$$\begin{aligned} \Lambda(k) &\triangleq \text{diag}\{\lambda_1(k), \cdots, \lambda_m(k)\}, \\ v(k) &\triangleq \text{vec}\{v_1(k), \cdots, v_m(k)\}, \\ y(k) &\triangleq \text{vec}\{y_1(k), \cdots, y_m(k)\}, \\ C &\triangleq \text{vec}\{C_1, \cdots, C_m\}. \end{aligned}$$

As such, the measurement equation can be written in a more compact form as follows

$$y(k) = \Lambda(k)Cx(k) + v(k). \tag{4.3}$$

For presentation convenience, we denote the statistics of $\Lambda(k)$ as

$$\begin{aligned} \bar{\Lambda} &\triangleq \mathbb{E}[\Lambda(k)] = \text{diag}\{\bar{\lambda}_1, \bar{\lambda}_2, \cdots, \bar{\lambda}_m\}, \\ \breve{\Lambda} &\triangleq \text{Var}[\Lambda(k)] = \text{diag}\{\breve{\lambda}_1, \breve{\lambda}_2, \cdots, \breve{\lambda}_m\}. \end{aligned}$$

The vector-valued nonlinear function $h(\cdot) : \mathbb{R}^n \to \mathbb{R}^n$ is analytic everywhere and satisfies the following condition

$$(h(x) - h(y))'(h(x) - h(y)) \leq \eta(x - y)'(x - y) \tag{4.4}$$

with $h(0) = 0$, where η is a positive scalar.

In this chapter, the networked state estimation problems are taken into account, i.e., the measurements $y_i(k)$ from every sensor are sent to the remote estimation center via communication channels. For certain applications, the network resource might be highly scarce or constrained, and it is, therefore, vitally important to reduce the sensor-to-estimator communication rate. To tackle this issue, we introduce an event-triggered communication strategy, under which the signals are transmitted if and only if the specific conditions are satisfied. Specifically, let us denote a time sequence that records the triggering instants of ith sensor by $\{k_s^i\}_{s=1}^{\infty}$, where

$$0 = k_1^i < k_2^i < \cdots < k_s^i < \cdots .$$

Suppose that the latest triggering instant of ith sensor is k_s^i, then the forthcoming triggering instant can be determined iteratively on-line by checking the triggering condition as follows:

$$k_{s+1}^i = \min_k\{k|k > k_s^i,\ |y_i(k) - y_i(k_s^i)| > \sigma_i\}, \tag{4.5}$$

where $\sigma_i > 0$ is a preassigned threshold, $y_i(k)$ and $y_i(k_s^i)$ represent the current and the latest transmitted measurements at the sampling step k, respectively.

From the event-triggered strategy, it can be seen that only when the inequality in (4.5) becomes true, the corresponding measurement will be transmitted to the remote estimator, otherwise it will be discarded. Between two successive event instants, the latest transmitted signal $y_i(k_s^i)$ is kept on the estimator by a zero-order holder. For simplicity, we define the latest triggered signal at sampling instant k by $\check{y}_i(k)$ (i.e., $\check{y}_i(k) = y_i(k_s^i)$, $\forall k \in [k_s^i, k_{s+1}^i)$) and denote $\check{y}(k) \triangleq \mathrm{vec}\{\check{y}_1(k), \check{y}_2(k), \cdots, \check{y}_m(k)\}$.

Next, we consider the triggered measurements $\check{y}_i(k)$ quantized before entering the network. The quantizer is defined by

$$\mathcal{Q}(\check{y}(k)) = \mathrm{vec}\{\mathcal{Q}_1(\check{y}_1(k)), \mathcal{Q}_2(\check{y}_2(k)), \cdots, \mathcal{Q}_m(\check{y}_m(k))\},$$

where $\mathcal{Q}_i(.)$ is assumed to be of the probabilistic uniform type. Specifically, the set of quantization levels for each $\mathcal{Q}_i(.)$ is described by

$$\mathcal{U}_i = \{\tau_{i,\kappa}|\tau_{i,\kappa} \triangleq \kappa\delta_i,\ \kappa = 0, \pm1, \pm2, \cdots\},\ \delta_i > 0.$$

The quantization function maps the whole segment to the quantization set $\mathcal{U}_i$. When $\tau_{i,\kappa} \le \check{y}_i(k) \le \tau_{i,\kappa+1}$, the signal $\check{y}_i(k)$ is quantized in the following probabilistic manner:

$$\begin{cases} \mathbb{P}\{\mathcal{Q}_i(\check{y}_i(k)) = \tau_{i,\kappa}\big|r_i\} = 1 - r_i \\ \mathbb{P}\{\mathcal{Q}_i(\check{y}_i(k)) = \tau_{i,\kappa+1}\big|r_i\} = r_i \end{cases} \tag{4.6}$$

where $r_i = (\check{y}_i(k) - \tau_{i,\kappa})/\delta_i \in [0, 1]$. As a result, the quantization error $e_i^q(k) = \mathcal{Q}_i(\check{y}_i(k)) - \check{y}_i(k)$ is a Bernoulli random variable taking values at $r_i\delta_i$ or $(r_i - 1)\delta_i$, i.e.,

$$\begin{cases} \mathbb{P}\{e_i^q(k) = -r_i\delta_i\big|r_i\} = 1 - r_i \\ \mathbb{P}\{e_i^q(k) = (1 - r_i)\delta_i\big|r_i\} = r_i \end{cases} \tag{4.7}$$

Since the quantization is carried out locally for every component, it can be proved that $\mathbb{E}[e_i^q(k)e_j^q(k)] = 0$, for $i \ne j$. Moreover, according to [20], the probabilistic uniform quantizer $\mathcal{Q}_i(.)$ is unbiased and the quantization error variance is upper bounded by $\delta_i^2/4$, that is,

$$\mathbb{E}[e_i^q(k)] = 0, \; \mathrm{Var}[e_i^q(k)] \leq \frac{\delta_i^2}{4}.$$

After receiving the quantized measurements, a Kalman-type filter of the following structure is performed at the remote estimator to extract the plant state:

$$\begin{aligned} \hat{x}(k|k-1) &= A\hat{x}(k-1|k-1) \\ \hat{x}(k|k) &= \hat{x}(k|k-1) - K_k(\mathcal{Q}(\check{y}(k)) - \bar{\Lambda}C\hat{x}(k|k-1)) \end{aligned} \tag{4.8}$$

where $\hat{x}(k|k-1)$ is the one-step prediction and $\hat{x}(k|k)$ is the updated estimate. The initial value is given as $\hat{x}(0|0) = 0$ and $K_k \in \mathbb{R}^{n\times m}$ is the filter gain matrix to be determined.

Remark 4.1 Compared with tradition periodic filters, the major difference with the adopted filter (4.8) lies in that the latest triggered measurement $\check{y}(k)$ is used as the quantizer input (rather than the current measurement $y(k)$). Specifically, if the change of the measurement stays within the event domain, i.e., $y_i(k) \in \{y_i(k) \in \mathbb{R} : |y_i(k) - \check{y}_i(k)| \leq \sigma_i\}$, we know that the current measurement is close enough to the latest transmitted one and, therefore, $\check{y}_i(k)$ is utilized to approximate the $y_i(k)$ as the quantizer input. Whenever an event is triggered at instant k, in view of (4.5), the latest transmitted signal trivially becomes $y_i(k)$, i.e., $\check{y}_i(k) = y_i(k)$, $\forall k \in \{k_s^i\}_{s=1}^{\infty}$. Thus, it is obvious that the inequality $|y_i(k) - \check{y}_i(k)| \leq \sigma_i$ is always true.

As discussed previously in the introduction, due to computational or implementation uncertainties in the hardware realizations, it is almost impossible to implement the filter gain K_k accurately. As such, the update stage of the filter algorithm (4.8) might suffer from the stochastic gain variation as follows:

$$\hat{x}(k|k) = \hat{x}(k|k-1) - (K_k + \Sigma_k)(\mathcal{Q}(\check{y}(k)) - \bar{\Lambda}C\hat{x}_{k|k-1}), \tag{4.9}$$

where Σ_k is the estimator gain variation. According to [19], the variation can be modeled by

$$\Sigma_k = \sum_{i=1}^{n_1} \beta_i(k)H_i,$$

where $H_i \in \mathbb{R}^{n\times m}$ are a set of known matrices and $\beta_i(k)$ is a zero-mean scalar-valued white noise with variance $\check{\beta}_i$. It is worth pointing out that, if not well handled, the gain variations are likely to dramatically decrease the filtering performance [17]. As such, the resilience properties shall be taken into account when designing the gain matrix K_k. Throughout the chapter, we assume that all the random variables, i.e., $w(k)$, $\alpha(k)$, $v_i(k)$, $\lambda_i(k)$, and $\beta_i(k)$, are white and mutually independent.

We denote the one-step prediction error, the estimation error and their corresponding covariances as

$$e_{k|k-1} \triangleq x(k) - \hat{x}(k|k-1), \ e_{k|k} \triangleq x(k) - \hat{x}(k|k),$$
$$P_{k|k-1} \triangleq E[e_{k|k-1}e'_{k|k-1}], \ P_{k|k} \triangleq E[e_{k|k}e'_{k|k}].$$

As will be clarified later, it is intractable to compute the exact covariance dynamics in an analytical form because of the existence of the quantization error as well as the event-induced error. Therefore, *the objective of this chapter* is to design and analyze the resilient filter that minimizes a tight upper bound on the actual estimation error covariances.

4.2 Main Results

In this section, we investigate the resilient filter design problem. First, denote the second moment of the state of the system (4.1) by $X(k) \triangleq \mathbb{E}[x(k)x'(k)]$.

Lemma 4.2.1 *The sequence of matrices $\{X(k)\}_{k=0}^{\infty}$ is always bounded by the solutions of the following recursive equation:*

$$\bar{X}(k+1) = A\bar{X}(k)A' + \eta\breve{\alpha}\mathrm{tr}(\bar{X}(k))I + Q \tag{4.10}$$

with the initial value $\bar{X}(0) = X(0)$, that is, $X(k) \leq \bar{X}(k)$, $\forall k \in \mathbb{N}$.

Proof It is not difficult to verify that

$$\begin{aligned} &\mathbb{E}[h(x(k))h(x(k))'] \leq \mathbb{E}[h(x(k))'h(x(k))]I \\ &\leq \eta\mathbb{E}[x(k)'x(k)]I = \eta\mathrm{tr}(\mathbb{E}[x(k)x(k)'])I, \end{aligned} \tag{4.11}$$

where the second inequality follows from (4.4). As a result, we have

$$X(k+1) \leq AX(k)A' + \eta\breve{\alpha}\mathrm{tr}(X(k))I + Q.$$

It is observed that the right-hand side of the above inequality is affine linear in $X(k)$, so $X(k+1) - \bar{X}(k+1) \leq 0$ if $X(k) \leq \bar{X}(k)$. Consequently, given the initial condition $\bar{X}(0) = X(0)$, $X(k) \leq \bar{X}(k)$ always holds.

Now, subtracting (4.8) from (4.1) results in the following dynamics of the one-step prediction error

$$e_{k+1|k} = Ae_{k|k} + \alpha(k)h(x(k)) + w(k).$$

By utilizing (4.11) and Lemma 4.2.1, we compute the error covariance as follows

$$P_{k+1|k} \leq AP_{k|k}A' + \eta\breve{\alpha}\mathrm{tr}(\bar{X}(k))I + Q. \tag{4.12}$$

In the light of the event-triggered communication rule (4.5), we can see that a time-varying gap exists between the current measurement $y(k)$ and the latest transmitted signal $\check{y}(k) \triangleq \text{vec}\{\check{y}_1(k), \check{y}_2(k), \cdots, \check{y}_m(k)\}$, denoted by $\check{e}(k) \triangleq \check{y}(k) - y(k)$ with the constraint $\|\check{e}(k)\|^2 \leq \sum_{i=1}^{m} \sigma_i^2 \triangleq \sigma$. By letting an augmented vector

$$e^q(k) \triangleq \text{vec}\{e_1^q(k), \cdots, e_m^q(k)\},$$

we can write the received measurement as follows

$$\mathcal{Q}(\check{y}(k)) = \check{y}(k) + e^q(k) = y(k) + \check{e}(k) + e^q(k). \tag{4.13}$$

Combining (4.1), (4.9), and (4.13), we obtain the dynamics of the estimate error as follows

$$\begin{aligned} e_{k|k} =&(I + (K_k + \Sigma_k)\bar{\Lambda}C)e_{k|k-1} + (K_k + \Sigma_k)(\check{e}(k) \\ &+ e^q(k) + \tilde{\Lambda}(k)Cx(k) + v(k)), \end{aligned}$$

where $\tilde{\Lambda}(k) = \Lambda(k) - \bar{\Lambda}$. The corresponding error covariance can be calculated as follows:

$$\begin{aligned} P_{k|k} =& \mathbb{E}[(I + (K_k + \Sigma_k)\bar{\Lambda}C)e_{k|k-1}e'_{k|k-1}(I + (K_k + \Sigma_k)\bar{\Lambda}C)'] \\ &+ \mathbb{E}[(K_k + \Sigma_k)\check{e}(k)\check{e}(k)'(K_k + \Sigma_k)'] + \mathbb{E}[(K_k + \Sigma_k)e^q(k)e^q(k)'(K_k + \Sigma_k)'] \\ &+ \mathbb{E}[(K_k + \Sigma_k)\tilde{\Lambda}(k)Cx(k)x(k)'C'\tilde{\Lambda}(k)(K_k + \Sigma_k)'] \\ &+ \mathbb{E}[(K_k + \Sigma_k)v(k)v(k)'(K_k + \Sigma_k)'] + \mathfrak{H} + \mathfrak{H}', \\ &+ \mathfrak{F} + \mathfrak{F}' + \mathfrak{G} + \mathfrak{G}' + \mathfrak{E} + \mathfrak{E}' \end{aligned} \tag{4.14}$$

where the cross terms are given by

$$\begin{aligned} \mathfrak{F} &= \mathbb{E}[(I + (K_k + \Sigma_k)\bar{\Lambda}C)e_{k|k-1}(\check{e}(k) + e^q(k) + \tilde{\Lambda}(k)Cx(k) + v(k))'(K_k + \Sigma_k)'] \\ \mathfrak{E} &= \mathbb{E}[(K_k + \Sigma_k)\check{e}(k)(e^q(k) + \tilde{\Lambda}(k)Cx(k) + v(k))'(K_k + \Sigma_k)'], \\ \mathfrak{G} &= \mathbb{E}[(K_k + \Sigma_k)e^q(k)(\tilde{\Lambda}(k)Cx(k) + v(k))'(K_k + \Sigma_k)'], \\ \mathfrak{H} &= \mathbb{E}[(K_k + \Sigma_k)\tilde{\Lambda}(k)Cx(k)v'(k)(K_k + \Sigma_k)']. \end{aligned}$$

Remark 4.2 It can be observed from (4.14) that, due to the uncertainties induced by the quantization error $e^q(k)$ and the event-triggered error $\check{e}(k)$, the exact characterization of the error covariance $P_{k|k}$ becomes analytically impossible, let alone designing the gain matrix K_k to minimize the covariance. To tackle this issue, in the following, an alternative approach is developed to cast the resilient filter design problem into a new one that aims to minimize the tight upper bound of the error covariance $P_{k|k}$.

Noting that $\lambda_i(k)$ is independent of $v_i(k)$, we have $\mathfrak{H} = 0$. Also, since the quantization error $e^q(k)$ is independent of $\tilde{\Lambda}(k)$ and $v(k)$, the term $\mathfrak{G}$ vanishes. Moreover, we have

$$\mathfrak{E} = \mathbb{E}[(K_k + \Sigma_k)\check{e}(k)(\tilde{\Lambda}(k)Cx(k) + v(k))'(K_k + \Sigma_k)'].$$

By letting ϵ_1 be a positive scalar and recalling a basic inequality $xy' + yx' \leq \epsilon_1 xx' + \epsilon_1^{-1}yy'$, we can derive that

$$\begin{aligned}\mathfrak{E} + \mathfrak{E}' \leq\ & \mathbb{E}[\epsilon_1^{-1}(K_k + \Sigma_k)\check{e}(k)\check{e}(k)'(K_k + \Sigma_k)'] \\ & + \mathbb{E}[\epsilon_1(K_k + \Sigma_k)(\tilde{\Lambda}(k)Cx(k) + v(k))(\tilde{\Lambda}(k)Cx(k) \\ & + v(k))'(K_k + \Sigma_k)'].\end{aligned}$$

Furthermore, as $\mathbb{E}[e^q(k)e^T_{k|k-1}] = 0$ and $e_{k|k-1}$ is independent of $\tilde{\Lambda}(k)$ and $v(k)$, it can be verified that

$$\mathfrak{F} = \mathbb{E}[(I + (K_k + \Sigma_k)\bar{\Lambda}C)e_{k|k-1}\check{e}(k)'(K_k + \Sigma_k)']$$

and, thus,

$$\begin{aligned}\mathfrak{F}+\mathfrak{F}' \leq\ & \epsilon_2^{-1}\mathbb{E}[(K_k + \Sigma_k)\check{e}(k)\check{e}(k)'(K_k + \Sigma_k)'] + \epsilon_2\mathbb{E}[(I \\ & + (K_k + \Sigma_k)\bar{\Lambda}C)P_{k|k-1}(I + (K_k + \Sigma_k)\bar{\Lambda}C)'].\end{aligned}$$

Substituting all the cross terms into (4.14) results in

$$\begin{aligned}P_{k|k} \leq\ & (1 + \epsilon_2)\mathbb{E}[(I + (K_k + \Sigma_k)\bar{\Lambda}C)P_{k|k-1}(I + (K_k \\ & + \Sigma_k)\bar{\Lambda}C)'] + \mathbb{E}[(K_k + \Sigma_k)e^q(k)e^q(k)'(K_k + \Sigma_k)'] \\ & + (1 + \epsilon_1)\mathbb{E}[(K_k + \Sigma_k)\tilde{\Lambda}(k)C\bar{X}(k)C'\tilde{\Lambda}(k)(K_k + \Sigma_k)] \\ & + (1 + \epsilon_1^{-1} + \epsilon_2^{-1})\mathbb{E}[(K_k + \Sigma_k)\check{e}(k)\check{e}(k)'(K_k + \Sigma_k)'] \\ & + (1 + \epsilon_1)\mathbb{E}[(K_k + \Sigma_k)v(k)v(k)'(K_k + \Sigma_k)'],\end{aligned}$$

where we have utilized the fact that $X(k) \leq \bar{X}(k)$. The first term on the right-hand side of the above inequality can be rearranged as follows

$$\begin{aligned}& \mathbb{E}[(I + (K_k + \Sigma_k)\bar{\Lambda}C)P_{k|k-1}(I + (K_k + \Sigma_k)\bar{\Lambda}C)'] \\ & = P_{k|k-1} + K_k\bar{\Lambda}CP_{k|k-1} + P_{k|k-1}C'\bar{\Lambda}K_k' \\ & \quad + \mathbb{E}[(K_k + \Sigma_k)\bar{\Lambda}CP_{k|k-1}\bar{\Lambda}C'(K_k + \Sigma_k)'],\end{aligned}$$

and

$$\begin{aligned}& \mathbb{E}[(K_k + \Sigma_k)\bar{\Lambda}CP_{k|k-1}\bar{\Lambda}C'(K_k + \Sigma_k)'] \\ & = K_k\bar{\Lambda}CP_{k|k-1} \times C'\bar{\Lambda}K_k + \sum_{i=1}^{n_1}\breve{\beta}_i H_i\bar{\Lambda}CP_{k|k-1}C'\bar{\Lambda}H_i'.\end{aligned}$$

According to the event-triggered rules, we have $\breve{e}(k)\breve{e}(k)' \leq \|\breve{e}(k)\|^2 I \leq \sigma I$. Since the variance of the quantized error is bounded, one can obtain that

$$\begin{aligned}\mathbb{E}[e^q(k)e^q(k)'] &= \text{diag}\left\{\mathbb{E}[|e_1^q(k)|^2], \cdots, \mathbb{E}[|e_m^q(k)|^2]\right\} \\ &\leq \text{diag}\{\delta_1^2/4, \cdots, \delta_m^2/4\} \triangleq \Delta.\end{aligned}$$

Denote $R \triangleq \text{diag}\{R_1, \cdots, R_m\}$ and

$$\Theta(\sigma, \bar{X}(k)) = \Delta + (1+\epsilon_1)R + (1+\epsilon_1)\breve{\Lambda} \circ C\bar{X}(k)C' + (1+\epsilon_1^{-1}+\epsilon_2^{-1})\sigma I.$$

It is straightforward to verify that

$$\begin{aligned}P_{k|k} &\leq (1+\epsilon_2)\left(P_{k|k-1} + K_k\bar{\Lambda}CP_{k|k-1} + P_{k|k-1}C'\bar{\Lambda}K_k'\right) \\ &+ K_k\mathcal{F}(\sigma, P_{k|k-1})K_k' + \sum_{i=1}^{n_1}\breve{\beta}_i H_i\mathcal{F}(\sigma, P_{k|k-1})H_i', \end{aligned} \tag{4.15}$$

where

$$\mathcal{F}(\sigma, P_{k|k-1}) = \Theta(\sigma, \bar{X}(k)) + (1+\epsilon_2)\bar{\Lambda}CP_{k|k-1}C'\bar{\Lambda}.$$

We construct two difference equations as follows:

$$M_{k|k-1} = AM_{k-1|k-1}A' + \eta\breve{\alpha}\text{tr}(\bar{X}(k))I + Q \tag{4.16}$$

$$\begin{aligned}M_{k|k} =& (1+\epsilon_2)\Big(M_{k|k-1} + K_k\bar{\Lambda}CM_{k|k-1} + M_{k|k-1}C'\bar{\Lambda}K_k'\Big) \\ &+ K_k\mathcal{F}(\sigma, M_{k|k-1})K_k' + \sum_{i=1}^{n_1}\breve{\beta}_i H_i\mathcal{F}(\sigma, M_{k|k-1})H_i' \end{aligned} \tag{4.17}$$

with the initial condition $M_{0|0} = P_{0|0}$. Now, we are ready to state our first main result in the following theorem.

Theorem 4.2.2 *The solutions of the Riccati-like recursions given in* (4.16) *and* (4.17) *are always the upper bounds of the actual error covariances, namely,* $P_{k+1|k} \leq M_{k+1|k}$ *and* $P_{k|k} \leq M_{k|k}$, $\forall\, k \in \mathbb{N}$.

Proof A straightforward algebraic manipulation shows that the recursion (4.17) is equivalent to the following in the quadratic form

$$\begin{aligned}M_{k|k} &= (1+\epsilon_2)(I + K_k\bar{\Lambda}C)M_{k|k-1}(I + K_k\bar{\Lambda}C)' \\ &+ \sum_{i=1}^{n_1}\breve{\beta}_i H_i\mathcal{F}(\sigma, M_{k|k-1})H_i' + K_k\Theta(\sigma, \bar{X}(k))K_k'. \end{aligned} \tag{4.18}$$

In what follows, an inductive method is exploited to prove our theorem. First, we know that $P_{0|0} = M_{0|0}$. Now, assume that $P_{k-1|k-1} \leq M_{k-1|k-1}$, $k \geq 1$. Subtracting

(4.16) from (4.12), we can derive $P_{k|k-1} - M_{k|k-1} \leq A(P_{k-1|k-1} - M_{k-1|k-1})A' \leq 0$, which implies $P_{k|k-1} \leq M_{k|k-1}$ and $\mathcal{F}(\sigma, P_{k|k-1}) - \mathcal{F}(\sigma, M_{k|k-1}) \leq 0$ because $\mathcal{F}(\sigma, P_{k|k-1})$ is an affine linear function of $P_{k|k-1}$. Furthermore, by combining (4.15) with (4.18), one has $P_{k|k} - M_{k|k} \leq 0$ which concludes that $P_{k|k} \leq M_{k|k}$, $\forall k \in \mathbb{N}$, which ends the proof.

The following theorem gives a solution to the filter design problem for minimizing the obtained upper bound on the actual error covariance.

Theorem 4.2.3 *The minimum upper bounds can be computed recursively by*

$$
\begin{aligned}
M_{k|k-1} &= AM_{k-1|k-1}A' + \eta\breve{\alpha}\mathrm{tr}(\bar{X}(k))I + Q \qquad (4.19)\\
M_{k|k} &= (1+\epsilon_2)M_{k|k-1} - K_k\mathcal{F}(\sigma, M_{k|k-1})K_k' \\
&\quad + \sum_{i=1}^{n_1} \breve{\beta}_i H_i \mathcal{F}(\sigma, M_{k|k-1})H_i' \qquad (4.20)
\end{aligned}
$$

with the initial condition $M_{0|0} = P_{0|0}$ *and the filter gain*

$$
K_k = -(1+\epsilon_2)M_{k|k-1}C'\bar{\Lambda}\left(\mathcal{F}(\sigma, M_{k|k-1})\right)^{-1}.
$$

Proof By completing the square with respect to the variable K_k, the following quadratic equation can be obtained

$$
\begin{aligned}
M_{k|k} =&(K_k - K_k^*)\mathcal{F}(\sigma, M_{k|k-1})(K_k - K_k^*)' \\
&+ (1+\epsilon_2)M_{k|k-1} - K_k^*\mathcal{F}(\sigma, M_{k|k-1})(K_k^*)' \\
&+ \sum_{i=1}^{n_1} \breve{\beta}_i H_i \mathcal{F}(\sigma, M_{k|k-1})H_i',
\end{aligned}
$$

where $K_k^* = -(1+\epsilon_2)M_{k|k-1}C'\bar{\Lambda}\left(\mathcal{F}(\sigma, M_{k|k-1})\right)^{-1}$. As $\mathcal{F}(\sigma, M_{k|k-1}) > 0$, by choosing the filter gain matrix $K_k = K_k^*$, we can obtain the minimized upper bound $M_{k|k}$ shown in this theorem. The proof is complete.

As discussed previously, the analytical expressions of the actual error covariance are unavailable due to the network-induced complexities. The solution of the difference Eqs. (4.19) and (4.20) actually provides a minimum upper bound of the actual error covariance and, therefore, serves as an alternative to reflect the state estimation accuracy. Moreover, When the errors $e^q(t)$ and $\breve{e}(t)$ and the nonlinear term $h(x(t))$ disappear, through slight modifications of the difference equations, it can be easily verified that such an upper bound reduces to the actual error covariance, see the following corollary.

Corollary 4.2.4 *When the errors* $e^q(t)$ *and* $\breve{e}(t)$ *and the nonlinear term* $h(x(t))$ *are zero, the minimum error covariance can be computed recursively by*

$$P_{k|k-1} = AP_{k-1|k-1}A' + Q$$

$$P_{k|k} = P_{k|k-1} - K_k\overline{\mathcal{F}}(P_{k|k-1})K_k' + \sum_{i=1}^{n_1} \breve{\beta}_i H_i \overline{\mathcal{F}}(P_{k|k-1})H_i'$$

with the filter gain $K_k = -P_{k|k-1}C'\bar{\Lambda}\left(\overline{\mathcal{F}}(P_{k|k-1})\right)^{-1}$, *where*

$$\mathcal{F}(P_{k|k-1}) = R + \breve{\Lambda} \circ \{CX(k)C'\} + \bar{\Lambda}CP_{k|k-1}C'\bar{\Lambda}.$$

Proof From (4.12) and (4.14), the error covariances can be rewritten as follows:

$$P_{k|k-1} = AP_{k-1|k-1}A' + Q$$

and

$$\begin{aligned}
P_{k|k} &= \mathbb{E}[(I + (K_k + \Sigma_k)\bar{\Lambda}C)e_{k|k-1}e'_{k|k-1}(I + (K_k + \Sigma_k)\bar{\Lambda}C)'] \\
&\quad + \mathbb{E}[(K_k + \Sigma_k)\tilde{\Lambda}(k)Cx(k)x(k)'C'\tilde{\Lambda}(k)(K_k + \Sigma_k)'] \\
&\quad + \mathbb{E}[(K_k + \Sigma_k)v(k)v(k)'(K_k + \Sigma_k)'] \\
&= P_{k|k-1} + K_k\bar{\Lambda}CP_{k|k-1} + P_{k|k-1}C'\bar{\Lambda}K_k' + K_k\overline{\mathcal{F}}(P_{k|k-1})K_k' \\
&\quad + \sum_{i=1}^{n_1} \breve{\beta}_i H_i \overline{\mathcal{F}}(P_{k|k-1})H_i'.
\end{aligned}$$

Moreover, the error covariance is minimized when the filter gain is chosen as follows

$$K_k = -P_{k|k-1}C'\bar{\Lambda}\left(\overline{\mathcal{F}}(P_{k|k-1})\right)^{-1}.$$

The resulting difference equation is

$$P_{k|k} = P_{k|k-1} - K_k\overline{\mathcal{F}}(P_{k|k-1})K_k' + \sum_{i=1}^{n_1} \breve{\beta}_i H_i \overline{\mathcal{F}}(P_{k|k-1})H_i'$$

which ends the proof.

Remark 4.3 To improve the tightness of the bound, it would be necessary to design the filter gain matrices to minimize the upper bound of error covariance at each step. We understand that it is almost impossible to make the bound arbitrarily tight for nonzero errors and the main conservatism results from the inequalities (17)–(18), but these inequalities are necessary in circumventing the complexity when dealing directly with the covariance between $\breve{e}(k)$ and the variables $e'_{k|k-1}$, $x(k)$ and $v(k)$.

Remark 4.4 Note that the minimum variance estimator has been intensively studied in the literature such as [21]. However, due to the errors $e^q(k)$ and $\breve{e}(k)$ in this chapter, the exact characterization of the error covariance $P_{k|k}$ becomes analytically impossible, let alone designing the gain matrix K_k to minimize the covariance. To handle this issue, an approximate minimum variance error filter with an analytical expression has been proposed in [4] by assuming distributions of the innovations are approximately Gaussian at each sampling instant. Nonetheless, such a filter is sensitive to the accuracy of density approximations and any deviation from the real probability density functions might lead to a great degradation of the estimation performance. Consequently, it is of significance to develop a novel filter under the minimum error variance framework while abandoning strict assumptions on density approximations.

4.3 Performance Analysis

In this section, we discuss the monotonicity of the minimum upper bound provided in Theorem 4.2.3 with respect to the event thresholds, and subsequently establish conditions for the boundedness of the steady-state error covariance.

4.3.1 Monotonicity

As shown in Theorem 4.2.3, the propagation of the covariance matrix depends on the event threshold σ. In this subsection, we will conduct a mathematically rigorous analysis on the monotonicity of the covariance with respect to the threshold. Intuitively, the increase of the thresholds might decrease the estimation accuracy since a larger change of measurements is required for triggering a transmission. Nonetheless, due to the complicated dynamics of (4.19) and (4.20), quantitative relations between the upper bound and the thresholds cannot be readily recognized, and there is a need to conduct a mathematically rigorous analysis on this issue.

At the very beginning, we denote a time-varying operator $\mathcal{G}_k(\cdot)$ as follows:

$$\begin{aligned}\mathcal{G}_k(\sigma, K, X) \triangleq\ & (1+\epsilon_2)A(I+K\bar{\Lambda}C)X(I+K\bar{\Lambda}C)'A' + Q \\ & + \sum_{i=1}^{n_1} \breve{\beta}_i A H_i \mathcal{F}(\sigma, X) H_i' A' + AK\Theta(\sigma, \bar{X}(k))K'A' \\ & + \eta\breve{\alpha}\mathrm{tr}(\bar{X}(k))I.\end{aligned}$$

where $\sigma = \sum_{i=1}^{m} \sigma_i^2$. It follows from (4.16) and (4.17) that the upper bound of one-step prediction error covariance can be written by $M_{k+1|k} = \mathcal{G}_k(\sigma, K_k, M_{k|k-1})$.

To facilitate the readability, in the following analysis, we will use the respective notations $M_{k|k}^{\sigma}$ and $M_{k+1|k}^{\sigma}$ to represent the minimized bounds $M_{k|k}$ and $M_{k+1|k}$ computed by solving the Riccati-like recursion (4.19) and (4.20) with the event threshold σ. The following proposition is concerned with the monotonicity.

Proposition 4.3.1 *Given two positive scalars $\bar{\sigma}_1$, $\bar{\sigma}_2$, if $\bar{\sigma}_1 \leq \bar{\sigma}_2$, then the following inequalities hold $M^{\bar{\sigma}_1}_{k+1|k} \leq M^{\bar{\sigma}_2}_{k+1|k}$, $\forall k \in \mathbb{N}$.*

Proof According to the proof in Theorem 4.2.3, M^{σ}_{k+1} can be minimized at step k if we choose the gain matrix $K^{\sigma,*}_k = -(1+\epsilon_2)M^{\sigma}_{k|k-1}C'\bar{\Lambda}(\mathcal{F}(\sigma, M^{\sigma}_{k|k-1}))^{-1}$, so

$$\mathcal{G}_k(\sigma, K^{\sigma,*}_k, M^{\sigma}_{k|k-1}) \leq \mathcal{G}_k(\sigma, K_k, M^{\sigma}_{k|k-1}), \forall K_k \in \mathbb{R}^{n\times m}.$$

Bearing this in mind, we will prove the monotonicity by exploiting an inductive method. For $k = 0$, the result is apparently true. Suppose that, at the sampling step $k-1 > 0$, the inequality $M^{\bar{\sigma}_1}_{k|k-1} \leq M^{\bar{\sigma}_2}_{k|k-1}$ holds. At the next sampling step, the minimum upper bound satisfies the following inequalities:

$$\begin{aligned} M^{\bar{\sigma}_1}_{k+1|k} &= \mathcal{G}_k(\bar{\sigma}_1, K^{\bar{\sigma}_1,*}_k, M^{\bar{\sigma}_1}_{k|k-1}) \\ &\leq \mathcal{G}_k(\bar{\sigma}_1, K^{\bar{\sigma}_2,*}_k, M^{\bar{\sigma}_1}_{k|k-1}) \\ &\leq \mathcal{G}_k(\bar{\sigma}_2, K^{\bar{\sigma}_2,*}_k, M^{\bar{\sigma}_2}_{k|k-1}) = M^{\bar{\sigma}_2}_{k+1|k}, \end{aligned}$$

where the second inequality follows from the monotonicity of the function $\mathcal{G}_k(\bar{\sigma}, K_k, X)$ in $\bar{\sigma}$ and X, and this ends the proof.

Remark 4.5 According to the Proposition 4.3.1, we can see the minimum upper bounds of the one-step prediction error covariance matrix (i.e., $M_{k+1|k}$) are monotonically increasing with respect to the event thresholds. This conclusion implies that there is a trade-off between the utilization rate of communication networks and the estimation accuracy. Moreover, it should be noted that the monotonicity of the upper bound with respect to the variance of sensor failure $\breve{\Lambda}$ as well as the bound of variance of the quantization error Δ can be proved in a similar way and the details are thus omitted for conciseness.

4.3.2 A Steady-State Filter

In Sect. 4.2, a resilient filter algorithm is established on a finite-horizon observation interval. However, it remains obscure whether the resilient filter converges or not as the interval tends to infinity. In addition, the realization of the proposed resilient filter requires the knowledge of the filter gain matrix K_k at every instant, and such a requirement is often difficult to achieve in real-time applications. Therefore, we are motivated to consider a steady-state filter by considering the case of $k \to \infty$ and investigate the corresponding convergence problem in this subsection.

Suppose that the sequence $\{\bar{X}(k)\}_{k=0}^{\infty}$ is always bounded by

$$\mathcal{X} \triangleq \min_{X}\{X \in \mathbb{R}^{n\times n}_{+} | \sup_{k\in\mathbb{N}} \bar{X}(k) \leq X\}.$$

An operator $\mathcal{L}(\cdot)$ can be introduced as follows

$$\begin{aligned}\mathcal{L}(\mathcal{M}) =&(1+\epsilon_2)A\mathcal{M}A' + (1+\epsilon_2)\sum_{i=1}^{n_1}\breve{\beta}_i AH_i\bar{\Lambda}C\mathcal{M}C'\bar{\Lambda}H_i'A' \\ &- (1+\epsilon_2)^2 A\mathcal{M}C'\bar{\Lambda}\left(\mathcal{F}(\sigma,\mathcal{M})\right)^{-1}\bar{\Lambda}C\mathcal{M}A'.\end{aligned}$$

With a slight abuse of notation, the first term in $\mathcal{F}(\sigma,\mathcal{M})$ is now replaced by $\Theta(\sigma,\mathcal{X})$ in the above operator, namely,

$$\mathcal{F}(\sigma,\mathcal{M}) = \Theta(\sigma,\mathcal{X}) + (1+\epsilon_2)\bar{\Lambda}C\mathcal{M}C'\bar{\Lambda}.$$

Denote

$$\Psi(X) \triangleq (1+\epsilon_2)\sum_{i=1}^{n_1}\breve{\beta}_i AH_i\Theta(\sigma,X)H_i'A' + \eta\breve{\alpha}\mathrm{tr}(X)I + Q.$$

The following theorem provides the result concerning the steady-state behavior of the proposed filter.

Theorem 4.3.2 *Suppose that* $\|A\|^2 + n\eta\breve{\alpha} < 1$ *and there exists a positive definite matrix* $\mathcal{M} > 0$ *such that* $\mathcal{M} > \mathcal{L}(\mathcal{M})$, *then* $\mathcal{P} = \mathcal{L}(\mathcal{P}) + \Psi(\mathcal{X})$ *has a unique symmetric positive definite solution. Moreover, when the steady-state filter gain is chosen as* $K = -(1+\epsilon_2)\mathcal{P}C'\Lambda(\mathcal{F}(\sigma,\mathcal{P}))^{-1}$, *we have* $\lim_{k\to\infty} M_{k+1|k} < \mathcal{P}$.

Proof Noting that the evolution of the sequence $\{M_{k+1|k}\}_{k=0}^{\infty}$ is coupled with $\bar{X}(k)$, it is obvious that if $\bar{X}(k)$ diverges as k tends to infinity then $M_{k+1|k}$ will also diverge. As such, at first, we need to establish conditions guaranteeing that the solution of $\bar{X}(k)$ is always bounded. Taking norms to the both side of the Eq. (4.10) yields

$$\|\bar{X}(k+1)\| \le \|A\|^2\|\bar{X}(k)\| + n\eta\breve{\alpha}\|\bar{X}(k)\| + \|Q\|,$$

which implies that, if $\|A\|^2 + n\eta\breve{\alpha} < 1$ is true, then there exists a positive definite matrix $\mathcal{X}$ such that $\bar{X}(k) \le \mathcal{X}$ always holds. Further, we define a new operator as follows

$$\begin{aligned}\mathcal{L}_1(\mathcal{M},\ K) =&(1+\epsilon_2)A(I+K\bar{\Lambda}C)\mathcal{M}(I+K\bar{\Lambda}C)'A' \\ &+ (1+\epsilon_2)\sum_{i=1}^{n_1}\breve{\beta}_i AH_i\bar{\Lambda}C\mathcal{M}C'\bar{\Lambda}H_i'A' + AK\Theta(\sigma,\mathcal{X})K'A'.\end{aligned}$$

It is not difficult to observe that $\mathcal{L}_1(\mathcal{M}, K)$ is a linear function of $\mathcal{M}$. Letting $\mathcal{M}_1 < \mathcal{M}_2$, we have

$$\mathcal{L}(\mathcal{M}_1) = \mathcal{L}_1(\mathcal{M}_1, K_1) \le \mathcal{L}_1(\mathcal{M}_1, K_2) \le \mathcal{L}_1(\mathcal{M}_2, K_2) = \mathcal{L}(\mathcal{M}_2),$$

where $K_i = \arg_K \min \mathcal{L}(\mathcal{M}_i, K), i = 1, 2$. Introduce a dynamic equation

$$\bar{M}_{k+1} = \mathcal{L}(\bar{M}_k) + \Psi(\mathcal{X}).$$

Obviously, when the conditions in this theorem are satisfied, there must exist a scalar $\mu_1 \in (0, 1)$ such that $\mathcal{L}(\mathcal{M}) \leq \mu_1 \mathcal{M}$. Moreover, we can choose a sufficiently large scalar $\mu_2 > 0$ in order to guarantee $\bar{M}_0 \leq \mu_2 \mathcal{M}$ and $\Psi(\mathcal{X}) \leq \mu_2 \mathcal{M}$. Noting that $\mathcal{L}(\mu_2 \mathcal{M}) < \mu_2 \mathcal{L}(\mathcal{M})$, we have

$$\begin{aligned}\bar{M}_1 &= \mathcal{L}(\bar{M}_0) + \Psi(\mathcal{X}) \leq \mathcal{L}(\mu_2 \mathcal{M}) + \mu_2 \mathcal{M} \\ &< \mu_2 \mathcal{L}(\mathcal{M}) + \mu_2 \mathcal{M} \leq \mu_1 \mu_2 \mathcal{M} + \mu_2 \mathcal{M}.\end{aligned}$$

By using the inductive method, it can be verified that

$$\bar{M}_k < \sum_{i=0}^{k} \mu_1^i \mu_2 \mathcal{M} < \frac{\mu_2}{1 - \mu_1} \mathcal{M},$$

which indicates that $\bar{M}_k$ is always bounded. When $\bar{M}_0$ is chosen to be $M_{1|0}$, we have

$$\begin{aligned}\bar{M}_1 &= \mathcal{L}(\bar{M}_0) + \Psi(\mathcal{X}) \geq \mathcal{L}(M_{1|0}) + \Psi(\bar{X}(0)) = M_{2|1}, \\ \bar{M}_2 &= \mathcal{L}(\bar{M}_1) + \Psi(\mathcal{X}) \geq \mathcal{L}(M_{2|1}) + \Psi(\bar{X}(1)) = M_{3|2}.\end{aligned}$$

By mathematical induction, we can see $\bar{M}_k$ is always the upper bound of $M_{k+1|k}$. Similar to Theorem 1 in [21], we can prove that, for any initial condition $\bar{M}_0$, $\bar{M}_{k+1} = \mathcal{L}(\bar{M}_k) + \Psi(\mathcal{X})$ converges, and its limit is independent of the initial condition. That is, $\mathcal{P} = \lim_{k\to\infty} \bar{M}_k$ exists and is the solution of the equation $\mathcal{P} = \mathcal{L}(\mathcal{P}) + \Psi(\mathcal{X})$. Therefore, we can draw the conclusion that $\lim_{k\to\infty} M_{k+1|k} < \lim_{k\to\infty} \bar{M}_k = \mathcal{P}$.

It is worth pointing out that, by exploiting the Schur Complement Lemma [22], the inequality $\mathcal{M} > \mathcal{L}(\mathcal{M})$ can be transformed into the linear matrix inequality (LMI) as follows:

$$\begin{pmatrix} (1+\epsilon_2)A\mathcal{M}A' - \mathcal{M} & \Pi_{12} & (1+\epsilon_2)A\mathcal{M}C'\bar{\Lambda} \\ * & -\Pi_{22} & 0 \\ * & * & -\mathcal{F}(\sigma, \mathcal{M}) \end{pmatrix} < 0,$$

where

$$\Pi_{12} = \left[(1+\epsilon_2)\breve{\beta}_1 A H_1 \bar{\Lambda} C \mathcal{M}, \cdots, (1+\epsilon_2)\breve{\beta}_m A H_m \bar{\Lambda} C \mathcal{M}\right],$$

and

$$\Pi_{22} = \operatorname{diag}\left\{(1+\epsilon_2)\breve{\beta}_1 \mathcal{M}, \cdots, (1+\epsilon_2)\breve{\beta}_m \mathcal{M}\right\}.$$

Consequently, a standard LMI technique can be utilized to verify whether the condition is satisfied in the above theorem.

Remark 4.6 The stability of the system matrix A is necessary for the convergence analysis of our filter to be carried out later. In case that both the random sensor failures and the stochastic nonlinearities are removed (i.e., $\lambda_i(k) \equiv 1$ and $\alpha(k) \equiv 0$), the bound $\bar{X}(k)$ will not appear in the error covariance and, subsequently, the system matrix A would no longer be restricted to be stable.

4.4 An Illustrative Example

In this section, a numerical example is presented to verify the effectiveness of the proposed resilient filter. We consider the linear discrete-time system (4.1) and (4.3) with the transition matrix and the observation matrix

$$A = \begin{pmatrix} -0.71 & 0.09 & -0.2 & -0.05 \\ 0.8 & 0 & 0 & 0 \\ 0 & 0.5 & 0 & 0 \\ 0 & 0 & 0.12 & 0 \end{pmatrix}, \quad C = \begin{pmatrix} 0 & 0 & 1 & 0 \\ 1 & 0 & 0 & 0 \end{pmatrix}.$$

The covariances of the process noise and measurement noise are given by $Q = 0.5I$ and $R = 0.3I$. The stochastic nonlinearity term is $h(x(k)) = |x(k)|$ and it is not hard to verify that $\eta = 1$. $\alpha(k)$ is a zero-mean random variable with variance $\breve{\alpha} = 0.01$. The initial state $x(0)$ is a Gaussian random vector with zero mean and covariance $X_0 = 100I$. The random sensor failure rates $\lambda_i(k)$ can be identified by statistical tests, which is assumed to obey the same probability density function $p(s) = 0.5X_1(s) + 0.2X_{0.5}(s) + 0.3X_0(s)$, where $X_a(\cdot)$ is an indicator function. Moreover, we set the quantization level $\delta_i = 0.1$ and the threshold $\sigma_i = 0.6$, for $i = 1, \ldots, m$. The random gain perturbation takes the form $\Sigma_k = \beta_1(k)H_1$, where

$$H_1 = \begin{pmatrix} 0 & 0.01 & -0.1 & 0.1 \\ 0.1 & 0.3 & 0 & 0.02 \end{pmatrix}'$$

and $\beta_1(k)$ is a zero-mean Gaussian variable with variance $\breve{\beta}_1 = 0.04$. The rest of the parameters are set to be $\epsilon_1 = 1$ and $\epsilon_2 = 1$. The purpose of the example is to compute the filter gains K_k such that, for all admissible perturbations and uncertainties, we can minimize an upper bound of the actual error covariance.

The simulation results are shown in Figs. 4.1, 4.2, and 4.3. From Fig. 4.1, we can see that by using the event-triggered scheme, only a small part of measurements are transmitted and, therefore, the utilization rate of the communication networks is reduced. Figure 4.2 depicts the upper bound and the actual mean-square error. Since the actual mean-square error versus time cannot be analytically computed, we obtain its value experimentally through 3000 independent repeated trials. Moreover, to show

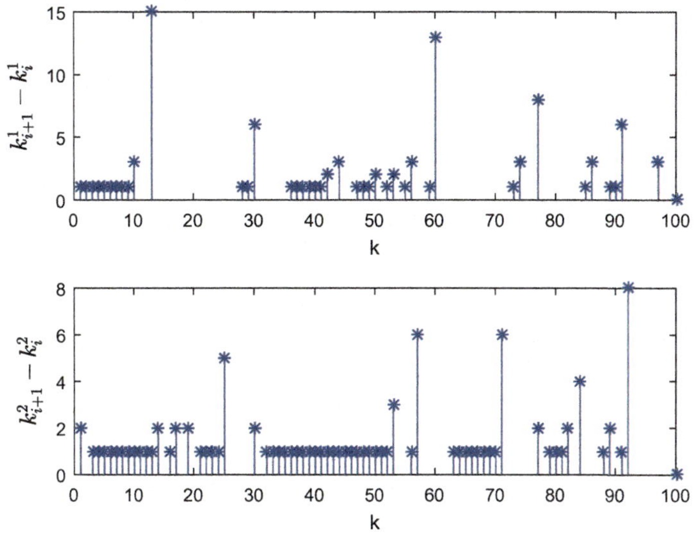

Fig. 4.1 The event-triggered instants and the intervals. The x-coordinate of the stems represents the instants when an event occurs and the length of the stems is the interval between two successive events

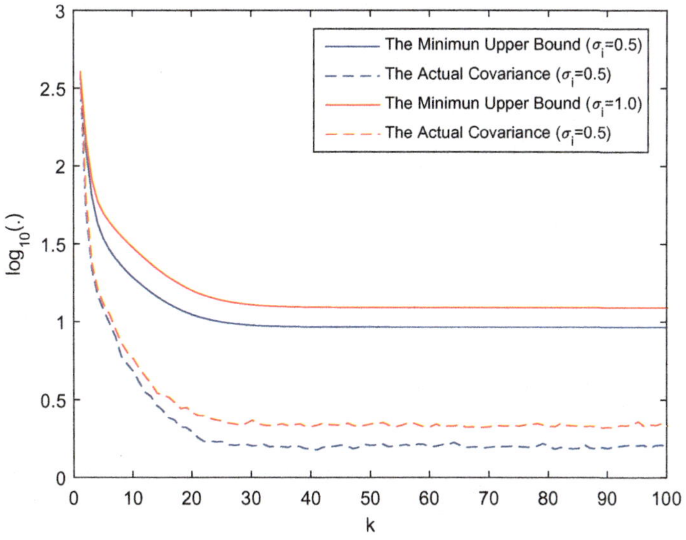

Fig. 4.2 Compare the upper bound $\log_{10}(\mathrm{tr}(M_{k+1|k}))$ and the mean-square errors $\log_{10}(\mathbb{E}[e'_{k|k} e_{k|k}])$ on the condition of different thresholds σ_i

the monotonicity of the minimum upper bound with respect to the threshold σ, we compare filtering performance with the different thresholds and present the result. As for the steady-state filter, it is obvious that the conditions in Theorem 4.3.2 are

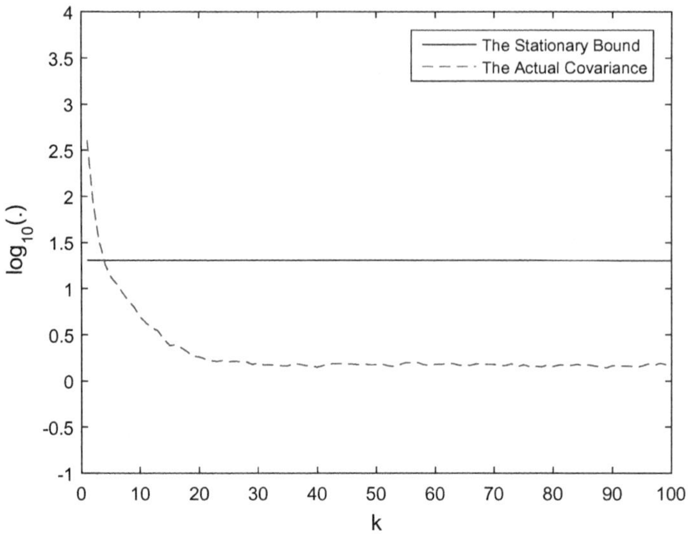

Fig. 4.3 Compare the steady-state bound $\log_{10}(\mathrm{tr}(\mathcal{P}))$ and the mean-square error $\log_{10}(\mathbb{E}[e'_{k+1|k} e_{k+1|k}])$

satisfied. By solving the matrix equality $\mathcal{P} = \mathcal{L}(\mathcal{P}) + \Psi(\mathcal{X})$, the steady-state gain matrix can be obtained. Subsequently, the actual mean-square error and the steady-state bound are depicted in Fig. 4.3. It can be shown that the steady-state mean-square error is upper bounded by $\mathrm{tr}(\mathcal{P})$.

4.5 Conclusions

In this chapter, we have addressed the resilient filter design problems for stochastic systems in the presence of random sensor failures and quantization effects in measurements. An event-triggered scheduling scheme has been introduced to reduce the communication rate through preventing the transmissions that are insignificant to enhance the estimate accuracy. Since the closed-form expressions of the error covariance cannot be directly derived in this scenario, we have resorted to its upper bound given in terms of Riccati-like recursions as an alternative. Subsequently, we have chosen the optimal filter gain matrix, such that the bound can be minimized recursively at each step. Further analysis has been carried out by discussing the monotonicity of the upper bound with respect to the thresholds and investigating the convergence of the steady-state filter. One of the future research topics would be the extension of our main results to the continuous-time systems, where the exclusion of the Zeno behavior becomes a critical issue.

References

1. W. Li, Y. Jia, J. Du, Distributed Kalman consensus filter with intermittent observations. J. Frankl. Inst. **352**(9), 3764–3781 (2015)
2. J. Linares-Perez, R. Caballero-Águila, I. Garcia-Garrido, Optimal linear filter design for systems with correlation in the measurement matrices and noises: recursive algorithm and applications. Int. J. Syst. Sci. **45**(7), 1548–1562 (2014)
3. J. Hespanha, P. Naghshtabrizi, Y. Xu, A survey of recent results in networked control systems. IEEE Proc. **95**(1), 138–162 (2007)
4. J. Sijs, M. Lazar, Event based state estimation with time synchronous updates. IEEE Trans. Autom. Control **57**(10), 2650–2655 (2012)
5. Q. Liu, Z. Wang, X. He, D.H. Zhou, Event-based H_∞ consensus control of multi-agent systems with relative output feedback: the finite-horizon case. IEEE Trans. Autom. Control **60**(9), 2553–2558 (2015)
6. X. Meng, T. Chen, Optimality and stability of event triggered consensus state estimation for wireless sensor networks, in *IEEE Proceedings of 2014 American Control Conference* (Portland, Oregon, 2014, pp. 3565–3570
7. W.P.M.H. Heemels, K.H. Johansson, P. Tabuada, An introduction to event-triggered and self-triggered control, in (Proceedings of IEEE Conference on Decision and Control, Maui, HI, 2012), pp. 3270–3285
8. M. Lemmon, Event-triggered feedback in control, estimation, and optimization, in *Networked Control Systems* (Springer, London, 2010), pp. 293–358
9. L. Li, M. Lemmon, Performance and average sampling period of sub-optimal triggering event in event triggered state estimation, in *2011 50th IEEE Conference on Decision and Control and European Control Conference*, pp. 1656–1661 (2011)
10. L. Li, B. Hu, M. Lemmon, Resilient event triggered systems with limited communication, in *2012 IEEE 51st Annual Conference on Decision and Control* (2012), pp. 6577–6582
11. L. Li, Z. Wang, M. Lemmon, Polynomial approximation of optimal event triggers for state estimation problems using SOSTOOLS, in *American Control Conference* (2013), pp. 2699–2704
12. S. Trimpe, R. D'Andrea, Event-based state estimation with variance-based triggering. IEEE Trans. Autom. Control **59**(12), 3266–3281 (2014)
13. N. Nahi, Optimal recursive estimation with uncertain observation. IEEE Trans. Inf. Theory **IT-15**(4), 457–462 (1969)
14. H. Leung, C. Seneviratne, M. Xu, A novel statistical model for distributed estimation in wireless sensor networks. IEEE Trans. Signal Process. **63**(12), 3154–3164 (2015)
15. M. Fu, C.E. de Souza, State estimation for linear discrete-time systems using quantized measurements. Automatica **45**(12), 2937–2945 (2009)
16. R. Zurkowski, S. Yuksel, T. Linder, On rates of convergence for markov chains under random time state dependent drift criteria. IEEE Trans. Autom. Control **161**, 145–155 (2016)
17. L.H. Keel, S.P. Bhattacharyya, Robust, fragile, or optimal? IEEE Trans. Autom. Control **42**(8), 1098–1105 (1997)
18. G. Yang, J. Wang, Non-fragile H_∞ control for linear systems with multiplicative controller gain variations. Automatica **37**(5), 727–737 (2001)
19. F. Hounkpevi, E. Yaz, Robust minimum variance linear state estimators for multiple sensors with different failure rates. Automatica **43**(7), 1274–1280 (2007)
20. T. C. Aysal, M. Coates, M. Rabbat, Rates of convergence for distributed average consensus with probabilistic quantization, in *Proceedings of the Allerton Conference on Communication, Control, and Computing* (2007)
21. B. Sinopoli, L. Schenato, M. Franceschetti, K. Poolla, M. Jordan, S. Sastry, Kalman filtering with intermittent observations. IEEE Trans. Autom. Control **49**(9), 1453–1464 (2004)
22. S. Boyd, L. Ghaoui, E. Feron, V. Balakrishnan, *Linear matrix inequalities in system and control theory* (SIAM, Philadelphia, 1994)

Chapter 5
Event-Based Distributed Filtering of Continuous-Time Nonlinear Systems

A wireless sensor network (WSN) is a distributed communication network containing an array of spatially separated intelligent sensing devices. Although each sensor node is typically small, power-constrained, low-cost, and sometimes unreliable, it can still perform various high-level tasks in a collaborative manner. As a consequence of the rapid development of microprocessors, wireless sensor networks have found a variety of potential applications ranging from environmental monitoring, traffic control, distributed robotics to integrated patient monitoring [1–5].

As one of the fundamental issues in wireless sensor networks, the distributed estimation problem has persistently attracted increasing research interest in recent years [6]. To be specific, in the distributed estimation setting, each sensor node is equipped with a filter that employs both the local and the neighboring information to obtain the true estimates of the observed states. Up to now, a large number of results have been reported in the literature. Generally speaking, there are mainly three approaches that shed insightful lights on the design and analysis of distributed filters among a rich body of literature. The first one is based on the minimum mean-square error (MMSE) criteria which aim at minimizing the error covariance, see e.g., [7, 8]. To be specific, in [7], a suboptimal filtering scheme has been considered to minimize the local quadratic filtering cost in each node, whereas in [8], the vector estimates from the decentralized sensors are allowed to be transmitted to a fusion center to obtain optimal estimation in the global sense. The main idea of the second approach is to find the stability conditions that guarantee the convergence of the filtering error dynamics by using Lyapunov theory, see e.g., [9]. The third method employs the H_∞ index to optimize worst-case filtering performance of the decentralized sensor networks with measurements contaminated by noises, see e.g., [10, 11].

In engineering practice, frequent data interaction via wireless links is likely to overspend the depletable bandwidth, and thus deteriorate communication quality. As a result, reducing the communication frequency between the sensors is essential

Q. Liu et al., *Stochastic Control and Filtering over Constrained Communication Networks*, Studies in Systems, Decision and Control 178,
https://doi.org/10.1007/978-3-030-00157-5_5

in prolonging working hours as well as guaranteeing the communication quality. During the past decades, the energy-efficient filtering methods have taken various forms in different studies on networked systems, such as interface-selection scheme [12], self-triggered communication [13], data rate constraint [14], and event-based transmission [15, 16]. To be more specific, a total communication rate constraint has been considered in [14] that can reduce the communication collision leading to energy waste. The main idea in [16] was to abandon the classical clock-driven mechanism and adopt an event-based approach where the communication was triggered only when some meaningful events occurred, thereby decreasing the number of transmissions through the channel.

In a typical event-based implementation, a transmission is triggered only when certain specified events occur. Different from the traditional clock-driven mechanism that predetermines periodic transmission instants based on the worst-case scenario to keep a desired system performance, the transmission instants of the event-based approach are automatically regulated online to selectively broadcast the significant data so as to reduce the number of transmissions to a large extent. Following the line of the work [16], the event-based filtering problems have been further investigated in several papers. In particular, the authors of [17, 18] have proposed a send-on-delta (SOD) concept scheduling aperiodic transmissions of sampling data to a remote Kalman filter for processing. In order to reduce the filtering error of the Kalman filters without real-valued innovations, a hybrid update strategy has been considered. Furthermore, the work [19] has considered a more general bounded Borel set in discrete-time measurement space to characterize the event domain, which is suitable for any type of event sampling. Then, a general mathematical description of event sampling has been presented by employing standard probability density functions. In addition, event-based H_∞ filtering problems have been investigated in [20] to suppress the influence of external disturbances on state estimation. By using linear matrix inequalities (LMIs), sufficient conditions have been established in [20] for the H_∞ performance.

On another research front, so far, considerable research efforts have been made on the filtering problems concerning the nonideal measurements that have been usually modeled by Bernoulli distribution. In such a distribution, the measurement signals from sensors are assumed to be either utterly missing or completely unchanged. This assumption is, unfortunately, not always true in the practical applications in the case of sensor networks due, primarily, to a variety of complicated internal and external factors such as sensors aging, sensor intermittent failure, high maneuverability of a tracked target as well as the influence from noisy environment. Therefore, to reflect the reality, it makes practical sense to consider the phenomena of stochastic measurement fading instead [21], where the probability function is governed by an individual random variable distributed over the interval [0, 1] with known statistical properties.

Summarizing the discussion above, it can be concluded that the event-based filtering problems have attracted some initial research interest focusing on the single filter only. When it comes to sensor networks, despite its importance in collaborative signal processing, the available results concerning distributed filtering problem have been scattered. Such a situation gives rise to the first motivation for our current research. Moreover, in the previous literature [17, 18], the SOD concept has been employed to characterize the distributed triggering rules, but the quantitative influence of the event threshold on the estimation error still remains uninvestigated which leads to the second motivation of our research. Furthermore, it is well recognized that the setting of the event threshold trades off the communication rate and estimation quality [22]. A large event threshold can reduce the number of transmissions at the cost of deteriorating the estimation quality. Therefore, in order to obtain satisfactory estimation quality, one would like to fine-tune the threshold so as to play the balance between the estimation accuracy and the resource consumption. Nevertheless, in most of the earlier studies, such a threshold is prefixed as a constant which cannot be tuned according to the real-time requirements. As such, the third motivation of our research is to develop an adaptive algorithm for determining the triggering threshold with hope to keep the average transmission rate level off a desired value.

The main contributions can be highlighted as follows. *(1) A novel event-based mechanism is proposed for the distributed filtering problem in order to reduce the amount of data transfer and save valuable resources in sensor networks. (2) The stochastic measurement fading phenomena are considered, thereby better characterizing the unreliability of individual sensors. (3) An adaptive threshold algorithm is designed for each sensor so as to obtain the desired average transmission rate according to the limitation of the wireless channel resources.*

5.1 Problem Formulation and Preliminaries

5.1.1 Wireless Sensor Networks

Consider the following nonlinear continuous-time Itô-type stochastic system (target plant):

$$dx(t) = f(x(t), t)dt + v(t)d\nu(t), \quad x(0) = x_0 \tag{5.1}$$

where $x(t) \in \mathbb{R}^n$ is the state vector to be estimated, $v(t) \in \mathbb{R}^n$ is an external noise intensity function, and $\nu(t)$ is a scalar Wiener process defined on probability space $(\Omega, \mathcal{F}, \mathcal{P})$ with $\mathbb{E}\{d\nu(t)\} = 0$ and $\mathbb{E}\{[d\nu(t)]^2\} = dt$.

In this chapter, we consider a wireless sensor network consisting of N sensor nodes, as shown in Fig. 5.1, to track the state of the target plant. These nodes are small in size but can perform some simple functions including, sensing, wireless communication, and signal processing. They share data with the adjacent nodes through wireless links to perform a certain task cooperatively, i.e., estimate target state. The directed interaction topology of a wireless sensor network is represented by a graph $\mathcal{G} = (V, E, A)$ with the set of nodes $\mathcal{V} = \{1, 2, \ldots, N\}$, set of edges $\mathcal{E} \subseteq V \times V$, and adjacency matrix $\mathcal{A} = [\alpha_{ij}]_{N \times N}$ with nonnegative elements α_{ij}. The edge $(i, j) \in \mathcal{E}$, if and only if, $\alpha_{ij} > 0$, which represents that the ith node can

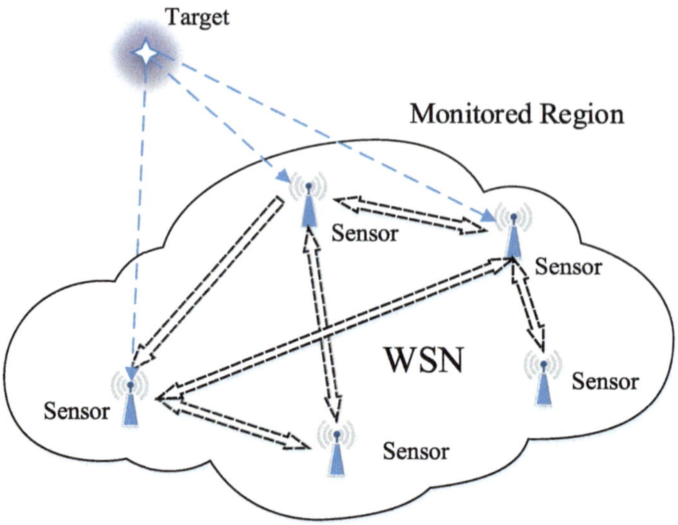

Fig. 5.1 The architecture of the WSN

receive the information from the jth node, otherwise, $\alpha_{ij} = 0$. Furthermore, self-loops are not allowed here, i.e., $\alpha_{ii} = 0$, for $i = 1, 2, \ldots, N$. All the nodes that could directly transmit the information to the ith node make up the set of neighbors as $\mathcal{N}_i \triangleq \{j \in \mathcal{V} | (i, j) \in \mathcal{E}\}$. The graph is assumed to be strongly connected, that is, there is at least a directed path from every node to every other node.

The model of ith sensor is described by

$$y_i(t)dt = x(t)dt + w_i(t)d\zeta_i(t) \tag{5.2}$$

where $y_i(t) \in \mathbb{R}^n$ is the measurement signal from ith sensor node, $w_i(t) \in \mathbb{R}^n$ is the intensity function of the external noise, and $\zeta_i(t) \in \mathbb{R}$ represents an independent Wiener process defined on probability space $(\Omega, \mathcal{F}, \mathcal{P})$ with $\mathbb{E}\{d\zeta_i(t)\} = 0$ and $\mathbb{E}\{[d\zeta_i(t)]^2\} = dt$.

The following assumptions are made on the target plant (5.1) and the measurement output (5.2).

Assumption 5.1 The vector-valued nonlinear function $f : \mathbb{R}^n \to \mathbb{R}^n$ is assumed to satisfy the following inequality:

$$(x - y)^T P(f(x, t) - f(y, t) - \Delta(x - y)) \leq -\eta(x - y)^T (x - y) \tag{5.3}$$

for $\forall x, y \in \mathbb{R}^n$, where η is a positive scalar, $P = \text{diag}\{p_1, p_2, \ldots, p_n\}$, and $\Delta = \text{diag}\{\delta_1, \delta_2, \ldots, \delta_n\}$.

Assumption 5.2 The components of the external noise intensity functions $v(t) = (v_1(t), v_2(t), \cdots, v_n(t))^T \in \mathbb{R}^n$ and $w_i(t) = (w_{i1}(t), w_{i2}(t), \cdots, w_{in}(t))^T$ satisfy the following inequalities

$$
\begin{aligned}
0 < v_k(t) \le \bar{v}_k \\
0 < w_{ik}(t) \le \bar{w}_{ik}
\end{aligned} \tag{5.4}
$$

for $i = 1, 2, \dots, N$ and $k = 1, 2, \dots, n$, where $\bar{v}_k$ and $\bar{w}_{ik}$ are positive constants.

Assumption 5.3 The random variables $\nu(t)$ and $\zeta_i(t)$ are mutually independent.

Remark 5.1 The nonlinear function $f(x)$, which contributes to the dynamical evolution of the target plant, satisfies the inequality shown in Assumption 5.1. Such an inequality is imposed on the changing rate of the nonlinear function so that the filter design problem is feasible. While Assumption 5.1 facilitates our theoretical development, it does make practical sense because the target states should not change too quickly in order to be traceable. Assumption 5.2 means that each component of the noise intensity function under consideration is bounded by a positive constant. Assumption 5.3 indicates the unrelatedness among the sources for stochasticity, namely, measurement noises and process noises.

5.1.2 Event-Based Distributed Filtering Strategies

As mentioned before, the sensor devices in wireless networks have independent data-processing and communication capability. To implement the distributed state estimation, the sensors not only utilize their local information but also need to interact with adjacent sensors. Therefore, based on the measurements $y_i(t)$ and the adjacent estimates $\hat{x}_j(t)$, $j \in \mathcal{N}_i$, the following local filters are adopted to restore the signal:

$$
\begin{aligned}
\frac{d\hat{x}_i(t)}{dt} =& f(\hat{x}_i(t), t) + c \sum_{j \in \mathcal{N}_i} \alpha_{ij}(\hat{x}_j(t) - \hat{x}_i(t)) \\
&+ c l_i(y_i(t) - \bar{\gamma}_i \hat{x}_i(t))
\end{aligned} \tag{5.5}
$$

for $i = 1, 2, \dots, N$, where

$$
\hat{x}_i(t) \triangleq (\hat{x}_{i1}(t), \hat{x}_{i2}(t), \cdots, \hat{x}_{in}(t))^T \in \mathbb{R}^n
$$

denotes the estimate of the state of the target in sensor i, c is the coupling strength between sensors, and l_i are positive scalars waited to be designed.

Note that the filters proposed are in the continuous-time setting. For practical applications, they should be implemented on digital devices via discretization. Traditional communication strategies rely on the time-scheduled mechanism. In other words, local estimates are broadcast to the adjacent nodes at each sampling step. However, such strategies are quite conservative in the sense that redundant information may be transmitted along with those vital data without any identification.

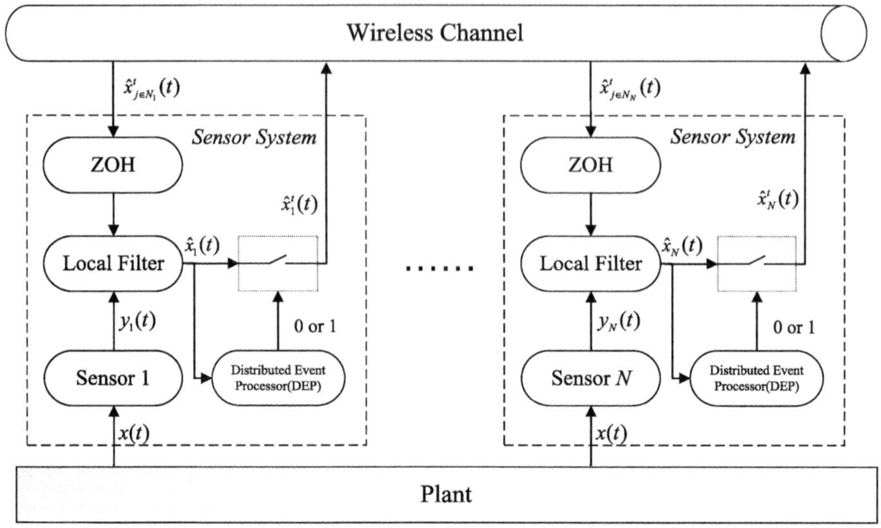

Fig. 5.2 Configuration of the event-based distributed filtering system

Moreover, in ad hoc wireless sensor networks subject to bandwidth and energy constraints, unnecessarily frequent signal transmission is not favorable because it will cut down the operating lifespan of distributed devices or even result in some negative network-induced phenomena, e.g., packet losses or transmission delays.

A possible way for overcoming the shortcoming of overspending communication resources is to specifically take the resource constraints into account in the filter design by defining the "valuable" messages sent via the wireless channel while ensuring the specified filtering performance. For this purpose, we reformulate the above time-scheduled filtering strategy and present a new transmission schedule called event-based mechanism, as shown in Fig. 5.2. The key modification is to replace $\hat{x}_j(t)$, $j \in \mathcal{N}_i$ in the filter (5.5) by piecewise signals. To be more formal, let us introduce the event instants for ith node as follows:

$$t_0^i = 0, t_1^i, \cdots, t_k^i, \cdots,$$

where $\{t_k^i > 0\}_{k=1}^{\infty}$ is a monotonically increasing sequence.

In order to determine the event instants iteratively online, the triggering rules ought to be established for ith sensor in Distributed Event Processor (DEP) as follows:

$$t_{k+1}^i = \inf\{t \in \mathbb{R}^+ | t > t_k^i, g_i(\hat{x}_i(t), \hat{x}_i(t_k^i), t) > 0\} \tag{5.6}$$

where $\mathbb{R}^+$ is a set of nonnegative real numbers. When $t \in [t_k^i, t_{k+1}^i)$, the event function $g_i(.) : \mathbb{R}^n \times \mathbb{R}^n \times \mathbb{R}^+ \to \mathbb{R}$ is given by

$$
\begin{aligned}
&g_i(\hat{x}_i(t), \hat{x}_i(t_k^i), \varphi_i) \\
&= (\hat{x}_i(t) - \hat{x}_i(t_k^i))^T P(\hat{x}_i(t) - \hat{x}_i(t_k^i)) - \varphi_i
\end{aligned} \tag{5.7}
$$

for $i = 1, 2, \ldots, N$. Therefore, whenever $g(\hat{x}_i(t), \hat{x}_i^t(t), \varphi_i) > 0$ is satisfied, the latest event instant is determined and, at this time, the DEP sends the permission signal "1" to trigger one transmission. Note that, in view of Fig. 5.2, we have implemented Zero-Order Holders (ZOHs) for all the nodes. Any received signal will be held constant via ZOHs separately, which is said to be "available" for the local node. In other words, every sensor is only aware of the estimates of the neighboring sensors at each latest triggering instant.

To present the transmission signals in a simple way, we take $\hat{x}_i^t(t)$ as the value of the estimate of sensor i at the latest event instant. For $t \in [t_k^i, t_{k+1}^i)$, it is easy to see that $\hat{x}_i^t(t) = \hat{x}_i(t_k^i)$. As a result, (5.5) could be modified into the following event-based filter

$$
\begin{aligned}
\frac{d\hat{x}_i(t)}{dt} =& f(\hat{x}_i(t), t) + c \sum_{j \in \mathcal{N}_i} \alpha_{ij}(\hat{x}_j^t(t) - \hat{x}_i^t(t)) \\
&+ cl_i(y_i(t) - \hat{x}_i(t)), \quad \text{for} \quad i = 1, 2, \ldots, N.
\end{aligned} \tag{5.8}
$$

In general, the scenario for the proposed event-based mechanism can be illustrated by taking the sensor i as an example. At the beginning, i.e., $t = 0$, the first event instant t_0^i is set to be 0, which indicates that sensors shall broadcast their signals to initialize sensor networks. During the filtering process, the ith sensor implements the algorithm (5.8) by making use of the local measurement $y_i(t)$ as well as the available signals $\hat{x}_j^t(t)$ to generate a real-valued estimation $\hat{x}_i(t)$. Meanwhile, the DEP embedded in sensor i checks the triggering rules (5.6) in order to determine the next event instants for triggering a transmission iteratively online. Repeating the above procedure every time, the event-based distributed filtering strategy can be implemented.

Define the coupling configuration matrix $A = [a_{ij}]_{N \times N}$ with the corresponding elements $a_{ij} = \alpha_{ij}$ for $i \neq j$, and $a_{ii} = -\sum_{i=1, i \neq j}^{N} a_{ij}$. Then, we have

$$
\frac{d\hat{x}_i(t)}{dt} = f(\hat{x}_i(t), t) + c \sum_{j=1}^{N} a_{ij} \hat{x}_j^t(t) + cl_i(y_i(t) - \hat{x}_i(t)) \tag{5.9}
$$

Subsequently, defining the local filter errors $e_i(t) = \hat{x}_i(t) - x(t)$ and denoting the event errors $e_i^t(t) = \hat{x}_i^t(t) - \hat{x}_i(t)$, it follows from (5.1) and (5.9) that

$$de_i(t) = \Bigg[f(\hat{x}_i(t), t) - f(x_i(t), t) + c\sum_{j=1}^{N} b_{ij} e_j(t) + c\sum_{j=1}^{N} a_{ij} e_j^t(t) \Bigg] dt + cl_i w_i(t) d\zeta_i(t) - v(t) d\nu(t), \tag{5.10}$$

where b_{ij} are the elements of the matrix $B \triangleq A - \text{diag}\{l_1, l_2, \cdots, l_N\}$.

Before proceeding, the following definition is introduced for characterizing the exponential mean-square stability.

Definition 5.1.1 The event-based distributed filtering error dynamics governed by (5.10) is said to be globally exponentially mean-square stable if there exist positive definite matrices Θ_i and three constants $\mu > 0$, $\tau > 0$, and $\varpi > 0$ such that

$$\sum_{i=1}^{N} \mathbb{E}\{\|e_i(t)\|_{\Theta_i}^2\} \leq \mu e^{-\tau t} \sum_{i=1}^{N} \mathbb{E}\{\|e_i(0)\|_{\Theta_i}^2\} + \varpi \tag{5.11}$$

We are now in a position to conclude the event-based distributed filtering problem as follows. The primary objective of this chapter is to design the filter gains l_i for each sensor such that the distributed filtering error (5.10) is exponentially mean-square stable under event-based mechanism.

5.2 Main Results

5.2.1 *Stability Analysis and Filter Design*

In what follows, a simple criterion for selecting the filtering gain will be given under the proposed setup so as to guarantee the exponential mean-square stability of the filtering error dynamics. Before establishing our main results, we need to introduce the following lemma to deal with the directed graph:

Lemma 5.2.1 ([23]) *Suppose that the directed graph $\mathcal{G}$ with the coupling configuration matrix A is strongly connected. Then, there exists a positive vector $\xi = (\xi_1, \xi_2, \cdots, \xi_N)^T$ such that $\xi^T A = 0$. Furthermore, $(\Xi A)^s = (\Xi A + A^T \Xi)$ is a symmetric matrix with zero row sum, where $\Xi = \text{diag}\{\xi_1, \xi_2, \cdots, \xi_N\}$.*

Theorem 5.2.2 *Consider the event-based distributed filtering problem associated with systems* (5.1) *and* (5.2), *where the interaction topology is strongly connected. Under Assumptions* (5.1) *and* (5.3), *if the following condition*

$$(2\delta_k + 1)\Xi + c(\Xi B)^s \leq 0, \tag{5.12}$$

is satisfied for $k = 1, 2, \ldots, n$, *then the distributed filtering error dynamics* (5.10) *with the event-based transmission schedule* (5.6) *and* (5.7) *is exponentially mean-square stable.*

Proof We aim to prove Theorem 5.2.2 based on the Lyapunov stability theory. First, construct the following Lyapunov–Krasovskii function for the error dynamics:

$$V(t) = \sum_{i=1}^{N} \xi_i e_i^T(t) P e_i(t), \tag{5.13}$$

where the positive definite diagonal matrices Ξ and P have been defined in Lemma 5.2.1 and Assumption 5.1, respectively. Define the infinitesimal operator $\mathcal{L}$ of $V(t)$ as follows:

$$\mathcal{L}V(t) = \lim_{\Delta \to 0^+} \frac{1}{\Delta}\Big\{\mathbb{E}\{V(t+\Delta)|e_1(t), \cdots, e_N(t)\} - V(t)\Big\}$$

By Itô differential formula, the stochastic differential of $V(t)$ along the trajectory of system (5.10) can be calculated by

$$dV(t) = \mathcal{L}V(t)dt + 2\sum_{i=1}^{N} \xi_i e_i^T(t) P[-v(t)d\nu(t) + cl_i w_i(t) d\zeta_i(t)], \tag{5.14}$$

where the infinitesimal operator $\mathcal{L}$ of the stochastic process is given by

$$\begin{aligned}\mathcal{L}V(t) =& \sum_{i=1}^{N} \Big\{2\xi_i e_i^T(t) P\big[f(\hat{x}_i(t), t) - f(x_i(t), t) + c\sum_{j=1}^{N} b_{ij} e_j(t) + c\sum_{j=1}^{N} a_{ij} e_j^t(t)\big] \\ &+ \xi_i v^T(t) P v(t) + c^2 l_i^2 \xi_i w_i^T(t) P w_i(t)\Big\}.\end{aligned} \tag{5.15}$$

In view of Assumption 5.1, we have

$$e_i^T(t) P\big[f(\hat{x}_i(t), t) - f(x_i(t), t)\big] \le -\eta e_i^T(t) e_i(t) + e_i^T(t) P \Delta e_i(t).$$

Denoting $\tilde{e}_k(t) = (e_{1k}^T(t), e_{2k}^T(t), \cdots, e_{Nk}^T(t))^T$, it can be verified that

$$\begin{aligned}
&2\sum_{i=1}^{N}\xi_i e_i^T(t)P\Delta e_i(t) &(5.16)\\
&=2\sum_{i=1}^{N}\sum_{k=1}^{n}\xi_i e_{ik}^T p_k\delta_k e_{ik}\\
&=2\sum_{k=1}^{n}p_k\delta_k\tilde{e}_k^T(t)\Xi\tilde{e}_k(t). &(5.17)
\end{aligned}$$

Meanwhile, we have

$$\begin{aligned}
&2\sum_{i=1}^{N}\xi_i e_i^T(t)P\Big(c\sum_{j=1}^{N}b_{ij}e_j(t)\Big)\\
&=2c\sum_{i=1}^{N}\sum_{j=1}^{N}\sum_{k=1}^{n}\xi_i e_{ik}^T(t)p_k b_{ij}e_{jk}(t)\\
&=c\sum_{k=1}^{n}p_k\tilde{e}_k^T(t)(\Xi B)^s\tilde{e}_k(t). &(5.18)
\end{aligned}$$

According to Assumption 5.2, one further obtains

$$\begin{aligned}
&\sum_{i=1}^{N}\xi_i v^T(t)Pv(t) &(5.19)\\
&=\sum_{i=1}^{N}\sum_{k=1}^{n}p_k\xi_i v_k^T(t)v_k(t)\\
&\le\sum_{i=1}^{N}\sum_{k=1}^{n}p_k\xi_i\bar{v}_k^2. &(5.20)
\end{aligned}$$

Similarly, we have

$$\sum_{i=1}^{N}c^2l_i^2\xi_i w_i^T(t)Pw_i(t)\le\sum_{i=1}^{N}\sum_{k=1}^{n}c^2l_i^2p_k\xi_i\bar{w}_{ik}^2. \tag{5.21}$$

Substituting (5.17)–(5.21) into (5.15) yields

$$
\begin{aligned}
\mathcal{L}V(t) \leq & -2\eta \sum_{i=1}^{N} \xi_i e_i^T(t) e_i(t) \\
& + \sum_{k=1}^{n} p_k \tilde{e}_k^T(t) \Big(\Xi(\delta_k I_N + cB) \Big)^s \tilde{e}_k(t) \\
& + 2 \sum_{i=1}^{N} \xi_i e_i^T(t) P \Big(c \sum_{j=1}^{N} a_{ij} e_j^t(t) \Big) \\
& + \sum_{i=1}^{N} \sum_{k=1}^{n} c^2 l_i^2 p_k \xi_i \bar{w}_{ik}^2 + \sum_{i=1}^{N} \sum_{k=1}^{n} p_k \xi_i \bar{v}_k^2.
\end{aligned} \tag{5.22}
$$

Moreover, based on the elementary inequality $x^T y + y^T x \leq x^T x + y^T y$, it can be concluded that

$$
\begin{aligned}
2 \sum_{i=1}^{N} \xi_i e_i^T(t) P \Big(c \sum_{j=1}^{N} a_{ij} e_j^t(t) \Big) \leq & \sum_{i=1}^{N} \xi_i e_i^T(t) P e_i(t) \\
& + \sum_{i=1}^{N} \xi_i \Big(c \sum_{j=1}^{N} a_{ij} e_j^t(t) \Big)^T P \Big(c \sum_{j=1}^{N} a_{ij} e_j^t(t) \Big).
\end{aligned} \tag{5.23}
$$

Next, letting $e^t(t) = (e_1^{tT}(t), e_2^{tT}(t), \cdots, e_N^{tT}(t))^T$, we have

$$
\begin{aligned}
& \sum_{i=1}^{N} \xi_i \Big(c \sum_{j=1}^{N} a_{ij} e_j^t(t) \Big)^T P \Big(c \sum_{j=1}^{N} a_{ij} e_j^t(t) \Big) \\
& = c^2 \Big((A \otimes I_n) e^t(t) \Big)^T (\Xi \otimes P) \Big((A \otimes I_n) e^t(t) \Big) \\
& = c^2 e^{tT}(t) \Big((A^T \Xi A) \otimes P \Big) e^t(t) \\
& \leq c^2 \lambda_{\max}\{A^T A\} \sum_{i=1}^{N} \xi_i e_i^{tT}(t) P e_i^t(t).
\end{aligned} \tag{5.24}
$$

Furthermore, let the threshold φ_i be a positive scalar. Suppose that the inequalities $(\Xi(\delta_k I_N + cB))^s \leq -\Xi$ and $e_i^{tT}(t) P e_i^t(t) \leq \varphi_i$ hold. Then, it follows that

$$
\mathcal{L}V(t) \leq -2\eta \sum_{i=1}^{N} \xi_i e_i^T(t) P e_i(t) + \varpi, \tag{5.25}
$$

where $\varpi = \sum_{i=1}^{N} \sum_{k=1}^{n} p_k \xi_i \bar{v}_k^2 + \sum_{i=1}^{N} \sum_{k=1}^{n} c^2 l_i^2 p_k \xi_i \bar{w}_{ik}^2 + c^2 \lambda_{\max}\{A^T A\} \sum_{i=1}^{N} \xi_i \varphi_i$.

According to the event-based mechanism (5.6), we can see that whenever the triggering condition $e_i^{tT}(t) P e_i^t(t) - \varphi_i > 0$ is fulfilled, a corresponding event is triggered.

Moreover, at each event instant, the step-wise function $\hat{x}_i^t(t)$ is updated/replaced by the real-valued estimate $\hat{x}_i(t)$, thereby resetting the event error $e_i^t(t)$ to zero immediately. In this case, under the event-based transmission schedule (5.6) and event function (5.7), the inequality $e_i^{tT}(t)Pe_i^t(t) \leq \varphi_i$ is always true.

Taking the mathematical expectation of both sides of (5.14) and substituting (5.25), it is derived that

$$\frac{\mathbb{E}\{dV(t)\}}{dt} = \mathbb{E}\{\mathcal{L}V(t)\} \leq -2\eta \sum_{i=1}^{N} \mathbb{E}\{\xi_i e_i^T(t)e_i(t)\} + \varpi. \tag{5.26}$$

It follows readily from (5.13) that

$$-2\eta \sum_{i=1}^{N} \xi e_i^T(t)e_i(t) \leq -(2\eta/\rho)V(t). \tag{5.27}$$

Substituting (5.27) into (5.26), one has

$$\frac{\mathbb{E}\{dV(t)\}}{dt} \leq -(2\eta/\rho)\mathbb{E}\big\{V(t)\big\} + \varpi. \tag{5.28}$$

Based on the Comparison principle [24], it can be concluded that the following inequality holds:

$$\begin{aligned} \mathbb{E}\{V(t)\} &\leq e^{-2\eta t/\rho}\mathbb{E}\{V(0)\} + \frac{1 - e^{-2\eta t/\rho}}{2\eta/\rho}\varpi \\ &\leq e^{-2\eta t/\rho}\mathbb{E}\{V(0)\} + \frac{\rho}{2\eta}\varpi. \end{aligned} \tag{5.29}$$

In view of Definition 5.1.1, we can conclude that the event-based distributed filtering error dynamics is exponentially mean-square stable, which ends the proof.

Remark 5.2 The threshold φ_i denotes the size of an event domain. A large φ_i decreases the frequency of event triggering, but will result in the increase of error bound ϖ. Particularly, when the thresholds $\varphi_i = 0$, the event-based transmission strategy reduces to the classical clock-driven one.

To further facilitate the application potential of the sufficient conditions given in Theorem 5.2.2, we have the following corollary.

Corollary 5.2.3 *Consider the event-based distributed filtering problem associated with systems* (5.1) *and* (5.2), *where the interaction topology is strongly connected. Let* $\psi = \lambda_{\max}\{(B)^s\}$, $\xi^+ = \max_i \xi_i$, $\xi^- = \min_i \xi_i$, $\rho = \max_i p_i$, *and* $\rho^- = \min_i p_i$. *With Assumptions (5.1)–(5.3), the exponential stability in the mean-square sense of the distributed filtering error dynamics* (5.9) *under the event-based transmission schedule* (5.6) *and* (5.7) *is guaranteed if*

$$(2\delta_k + 1) + c\psi \leq 0, \text{ for } k = 1, 2, \ldots, n, \tag{5.30}$$

Moreover, the bound of the expectation of the estimate error can be given by

$$\lim_{t\to\infty} \sum_{i=1}^{N} \mathbb{E}\left\{\|e_i(t)\|^2\right\} \leq \frac{\varpi\rho}{2\eta\xi^-\rho^-}. \tag{5.31}$$

Proof Noting that Ξ is a diagonal positive definite matrix, (5.12) can be obtained directly from (5.30). From the definition of the Lyapunov–Krasovskii function (5.13) and the inequality (5.29), it follows that

$$\sum_{i=1}^{N} \mathbb{E}\left\{\|e_i(t)\|^2\right\} \leq \frac{\xi^+\rho}{\xi^-\rho^-} e^{-2\eta t/\rho} \mathbb{E}\{\|e(0)\|^2\} + \frac{\varpi\rho}{2\eta\xi^-\rho^-}, \tag{5.32}$$

Consequently, it is not difficult to verify that the expectation of global mean-square errors satisfies (5.32), which completes the proof of Corollary 5.2.3.

Remark 5.3 In Theorem 5.2.2 and Corollary 5.2.3, sufficient conditions are presented to ensure the mean-square stability of the distributed filtering error dynamics. Further analysis reveals that the mean-square stability is related to three basic elements: (1) the network topology $\mathcal{G}$ that determines the coupling configuration matrix A; (2) the coupling strength c; (3) the filter parameters l_i that can regionally configure the poles of $(\Xi B)^s$.

An undirected network is clearly a special case of the directed networks. As such, if only the undirected interaction topology is concerned in certain wireless sensor networks for practical applications, we can further obtain the following criteria according to Theorem 5.2.2.

Corollary 5.2.4 *Consider the event-based distributed filtering problem associated with systems* (5.1)–(5.2). *Suppose that the undirected interaction topology under consideration is connected. With Assumptions 5.1–5.3, the exponential mean-square stability of the distributed filtering error dynamics* (5.9) *under the event-based transmission schedule* (5.6)–(5.7) *is guaranteed if*

$$(2\delta_k + 1) + 2c\lambda_{\max}\{B\} \leq 0, \text{ for } k = 1, 2, \ldots, n, \tag{5.33}$$

Moreover, the bound of the expectation of the estimate error is given by

$$\lim_{t\to\infty} \sum_{i=1}^{N} \mathbb{E}\left\{\|e_i(t)\|^2\right\} \leq \frac{\varpi_1\rho}{2\eta\rho^-}, \tag{5.34}$$

where $\varpi_1 = N\sum_{k=1}^{n} p_k\bar{v}_k^2 + \sum_{i=1}^{N} c^2 l_i^2 \sum_{k=1}^{n} p_k\bar{w}_{ik}^2 + c^2\lambda_{\max}\{A^T A\}\sum_{i=1}^{N}\varphi_i,$

Proof For an undirected network, the coupling configuration matrix A is a symmetric matrix with both zero row sum and zero column sum. Obviously, we can verify that $\xi^T = \mathbf{1}^T$ is the left eigenvector of the matrix A corresponding to the eigenvalue 0. Therefore, (5.12) and (5.31) can be transferred into (5.33) and (5.34), respectively. The proof is complete.

5.2.2 Adaptive Thresholds

With a given threshold φ_i, the communication rate can be reduced at the cost of increasing the error bound. A remaining issue deserving further investigation is how to properly choose the value of the threshold according to the practical wireless channel. In this section, we remove the assumption that the threshold is static and consider how such thresholds can be dynamically adjusted to make the most of the available channel bandwidth.

The event-based strategy is proposed to reduce the transmission frequency, but the average triggering rate cannot be specifically designed for each node in advance because of the random nature of the filtering errors. Furthermore, with a static triggering threshold, it would be difficult for the designer to properly manage the fast and slow dynamics simultaneously. To be more specific, in order to achieve the most effective utilization of the resources, the sensor ought to broadcast as much as possible according to its available bandwidth. For a target with slow dynamics, the threshold should be reduced to guarantee the precision of estimation while, for a target with fast dynamics, the threshold needs to be increased to preserve the bandwidth. Consequently, an adaptive algorithm dynamically adjusting the thresholds is of practical significance to handle the conflicts between the performance requirements.

Denote the average triggering rate $n_i(t)$ for ith node over the sliding window $[t-T, t)$ as follows:

$$n_i(t) = \frac{1}{T}\int_{t-T}^{t} \sum_{k=0}^{m_i(t)} \delta(s - t_k^i)ds, \tag{5.35}$$

where $\delta(.)$ is the Kronecker delta function, T represents the length of the observation window, t_k^i are the event instants defined in (5.6), and $m_i(t) = \max\{k \in \mathbb{R}^+ | t - t_k^i > 0\}$ indicates the accumulated triggering times at time t. Note that $n_i(t)$ is a stochastic variable presenting the average cost of communication over the sliding window. During the process, we would like to keep $n_i(t)$ fluctuating around the corresponding desired value $\bar{n}_i$, which can be tested based on the wireless channel bandwidth. For this purpose, we propose the adaptive threshold algorithm in Table 5.1. Whenever the real-time average triggering rate $n_i(t)$ is smaller than $\bar{n}_i$, the threshold will be enlarged to increase the number of triggering, otherwise the threshold will be reduced. Another parameter ϵ scaling the size of dead zone is introduced in order to avoid too frequent adjustments of the threshold when $n_i(t)$ fluctuates around $\bar{n}_i$.

Table 5.1 Adaptive threshold algorithm

Algorithm	
Step 1	Set the length of the observe window T, a large initial threshold φ_i and the desirable average triggering rate $\bar{n}_i$ for each sensor; choose appropriate parameters $\kappa_1 \in (-\infty, 0)$, $\kappa_2 \in (0, +\infty)$, and a scalar $\epsilon > 0$
Step 2	Calculate the real-time values of $n_i(t)$ by using (5.35)
Step 3	If $n_i(t) < (1-\epsilon)\bar{n}_i$, then $\dot{\varphi}_i(t) = \kappa_1\varphi_i(t)$ else if $n_i(t) > (1+\epsilon)\bar{n}_i$, then $\dot{\varphi}_i(t) = \kappa_2\varphi_i(t)$ otherwise, $\dot{\varphi}_i(t)$ =0 end if

Remark 5.4 If the communication bandwidth is not a concern, $\bar{n}_i$ can be set sufficiently large. According to the algorithm laid out in Table 5.1, the thresholds will drop exponentially and tend to zero quickly. In this case, the event-based strategy is reduced to a classical time-schedule one.

Remark 5.5 With the adaptive threshold algorithm, the threshold grows exponentially corresponding to a large real-valued average triggering rate $n_i(t)$, and subsequently such a rate will be decreasing gradually. Therefore, one can conclude that the filtering dynamic errors would never exhibit a Zeno behavior. In other words, it is almost impossible to have an infinite number of events in any finite time in the trajectory, which is in agreement with the reality.

5.3 An Illustrative Example

In this section, the localization problem of a wheeled mobile robot is addressed in order to illustrate the effectiveness of the proposed event-based distributed filtering scheme. The prototype constructed for this purpose consists of a wheeled mobile robot with a signal transmitter. What we are interested in is the positioning problem of the wheeled mobile robot via a wireless sensor network.

The wireless sensor network is composed of five separated nodes. Each node equips measuring units receiving the transmitter's signal, such that it has the rough positional information of the moving wheeled mobile robot [25]. Moreover, the node shares data with other sensors within its communication radius via wireless links to achieve cooperative estimation. Such a communication radius is determined by actual transmission power of the node in practical applications, but here we assume it as a prior for simplification. The profile of the wireless sensor network is shown in Fig. 5.3. To assist the readers, we further depict the schematic structure of the communication topology in Fig. 5.4.

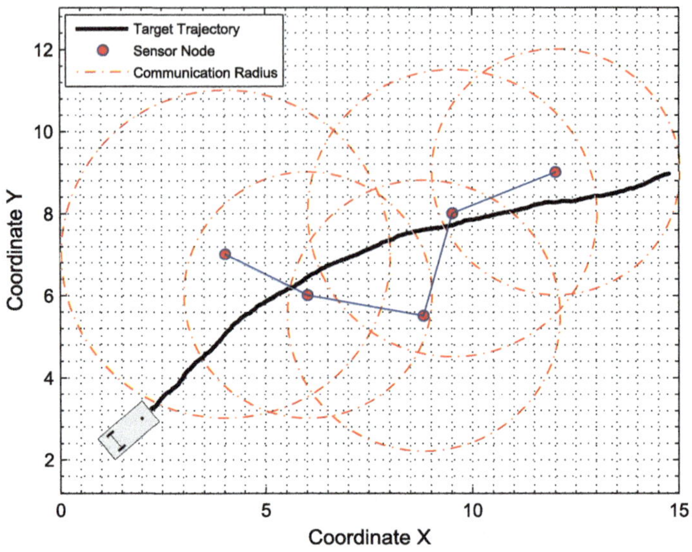

Fig. 5.3 The sensor network

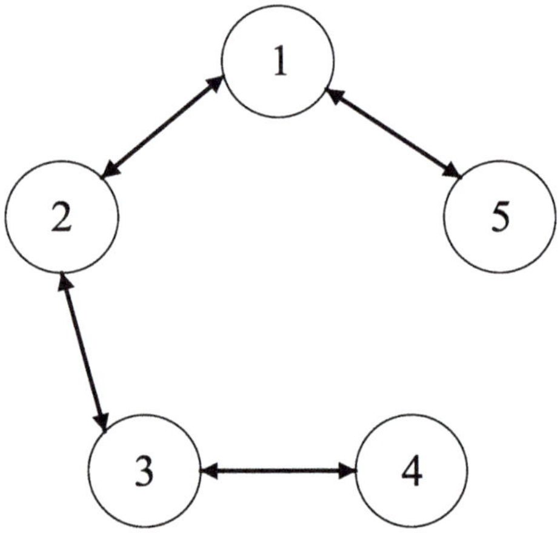

Fig. 5.4 The topology of the directed graph $\mathcal{G}$

Denote the unknown state of the wheeled mobile robot as $x = (p_x, p_y, v_x, v_y)^T$ with the initial state as $p_x(0) = 2$(m), $p_y(0) = 3$(m), $v_x(0) = 0.1$(m/s) and $v_y(0) = 0.1$(m/s), where p_x, p_y are position components and v_x, v_y are ground velocity components along the coordinate axes, respectively. For the purpose of simplicity, the trajectory of the wheeled mobile robot is assumed to be in the plane. We use the control scheme proposed in [26] to carry out point-to-point motion of the robot, whose actual dynamics is

$$f(x, t) = (v_x, v_y, 0.001, 0)^T.$$

The noise intensity functions $v(t)$ and $w_i(t)$ have been obtained by means of suitable preliminary experiments. In particular, we have $v_k(t) = 1.5$ and $w_{ik}(t) = 1$. Based on Assumption 5.1, one can derive that $P = \text{diag}\{1, 1, 1, 1\}$, $\Delta = \text{diag}\{0.51, 0.51, 0.51, 0.51\}$ and $\eta = 0.01$.

We run the following distributed filter algorithm for each sensor

$$\frac{d\hat{x}_i(t)}{dt} = f(\hat{x}_i(t), t) + c\sum_{j=1}^{N} a_{ij}\hat{x}_j^t(t) - cl_i(\hat{x}_i(t) - y_i(t)),$$

with the initial estimates $\hat{x}_i(0) = (0\,0\,0\,0)^T$ for $i = 1, 2, \ldots, 5$. The nonzero elements of the adjacent matrix $\mathcal{A}$ are designed as

$$\mathcal{A} = \begin{pmatrix} 0 & 0.12 & 0 & 0 & 0.61 \\ 0.54 & 0 & 0.67 & 0 & 0 \\ 0 & 0.64 & 0 & 0.23 & 0 \\ 0 & 0 & 0.37 & 0 & 0 \\ 0.93 & 0 & 0 & 0 & 0. \end{pmatrix} \tag{5.36}$$

Therefore, it follows from Lemma 5.2.1 that

$$\xi = [0.80, 0.18, 0.19, 0.12, 0.53]^T.$$

Choosing the coupling strength $c = 4$ and the filter gains $l_i = 2.4$ for $i = 1, 2, \ldots, 5$, the condition (5.12) in Theorem 1 is satisfied, and consequently, the distributed estimates $\hat{x}_i(t)$ will converge to the real state $x(t)$.

To achieve the objective of resource-efficient allocation, only the necessary data is transmitted during the filtering process. Further test of the wireless channel indicates that the best transmission rate for each sensor is average 9 times in 1 second. Therefore, the adaptive filter algorithm (Table 5.1) is applied to dynamically adjust the threshold so as to meet such a requirement. To be more specific, we choose the initial threshold as $\varphi_i = 3$, the length of the sliding window as $T = 0.6(s)$, and the desired average communication rate as $\bar{n}_i = 9$ per second for each node. The rest of the parameters are set to $\epsilon = 0.2$, $\kappa_1 = -10$, and $\kappa_2 = 10$. Moreover, to simulate the process of distributed filtering, the numerical simulation step dt is set as 0.001, and the event rules (5.6) are checked at each step.

The results are shown in Figs. 5.5, 5.6, and 5.7. According to Fig. 5.5, it can be seen that all the estimates of the sensor nodes will eventually converge to the real state. Note that the estimation error of the sensor 5 converges much slower than other nodes, because it has to estimate the state totally dependent on its adjacent nodes. In Fig. 5.6, the x-axis of the sticks gives the triggering instants, while the length of the sticks represents the inter-event time between two successive events. Figure 5.7 presents the adaptive adjustment of the thresholds. At the beginning of the process, the smart sensors recognize that the real-valued communication rate is less than the setting value, and therefore decrease the thresholds to shrink the event domain until

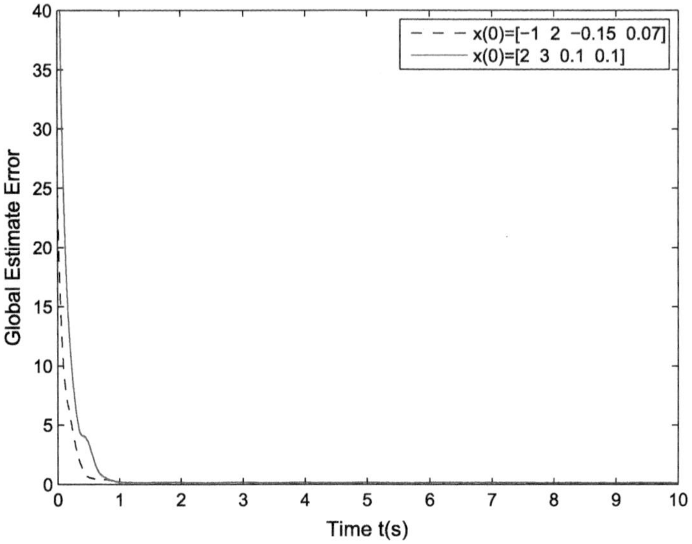

Fig. 5.5 Mean-square error of EBDF with initial target states x(0)

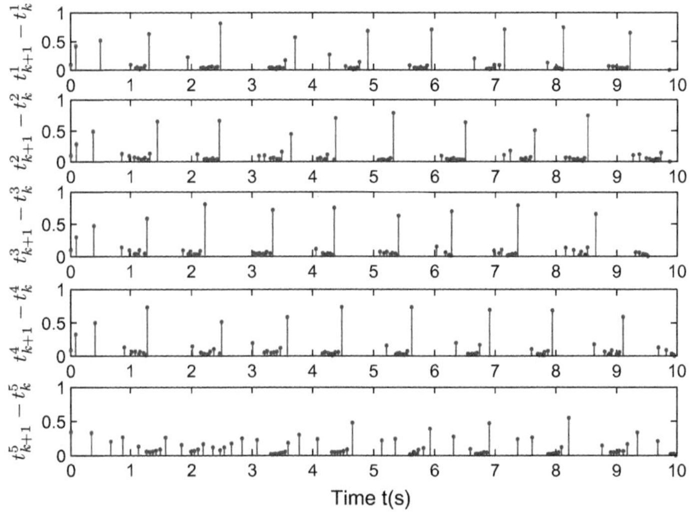

Fig. 5.6 The inter-event time for $k = 0, 1, 2, \ldots$

this rate is close to the setting one. Table 5.2 counts the number of event triggering throughout the process. We can find that in the total of 10, 000 sampling times, around 0.9% steps trigger the broadcast, so the average communication time is about 9 in every second, which meets the desired value quite well. Moreover, we repeat the experiment with various initial states $x(0)$, and the corresponding mean-square error

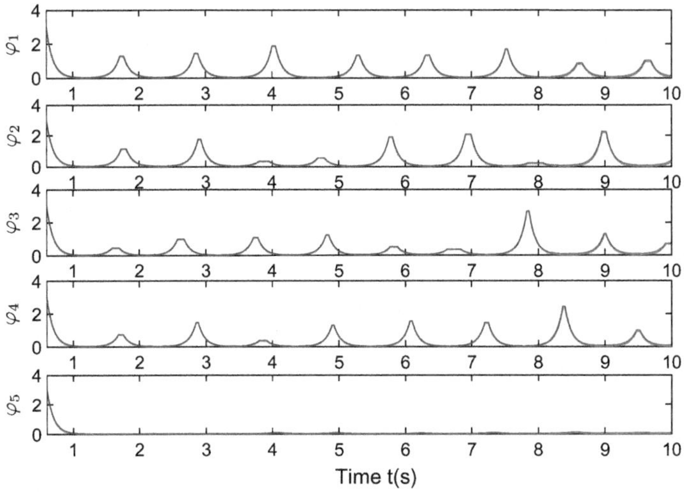

Fig. 5.7 The adaptive threshold

Table 5.2 Triggering times in total 20,000 steps

	Node 1	Node 2	Node 3	Node 4	Node 5
Total (20,000)	88	82	86	89	91
Percentage (%)	0.88	0.82	0.86	0.89	0.91

curves are plot in Fig. 5.5. Now, it can be seen that all of them show the exponential convergence property as expected.

5.3.1 Comparison with Consensus-Based Distributed Filters

In Fig. 5.8, the mean-square error of the proposed event-based distributed filter (EBDF) is compared with that of the consensus-based distributed filters(CBDF) in [7]. Note that CBDF is clock-driven, so the filter interacts with the neighboring nodes at each sampling instant. Although both of them achieve mean-square stability, EBDF converges a bit slower with a larger steady-state error. Nevertheless, this chapter is concerned with the issue of limited resource with focus on the energy-saving problem (rather than the pure accuracy issue). Since the EBDF has reduced the communication rate to only 0.9%, the current tradeoff between the communication cost and the accuracy is acceptable.

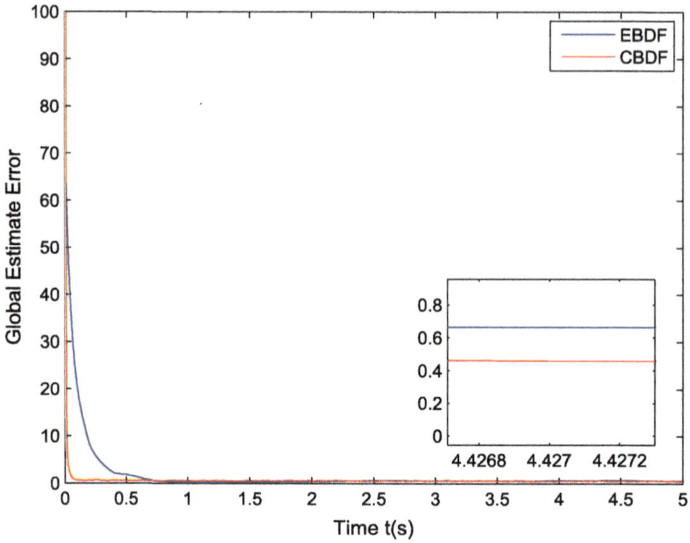

Fig. 5.8 EBDF compared with CBDF

5.4 Conclusions

In this chapter, the event-based distributed filtering problems have been investigated for a class of continuous-time Itô stochastic systems in the wireless sensor networks subject to finite resources and stochastic measurement fading. According to the event-based mechanism, the intelligent sensor cuts down the communication frequency by selectively transmitting those important data to its neighbors when the predetermined event rules are satisfied. A sufficient condition, which is linked to the design of filter gains and event domains, has been obtained such that the corresponding filtering error system is exponentially stable in the mean-square sense. Subsequently, an adaptive threshold algorithm has been proposed to maintain a desired average transmission rate during the system process. Finally, a tracking problem of the wheeled mobile robot has been taken into account in the numerical example to demonstrate the effectiveness and applicability of the developed strategy.

It is worth pointing out that the distributed estimation problem subject to a time-varying topology is an interesting and significant topic. Nevertheless, due to the time-varying nature of the system, a novel methodology would have to be developed, which would be one of our future research topics. Also, although the event-based filters reduce the communication rate, they may suffer from the loss of the estimation performance. The next step would be to analyze and design event-based filters in the guide of an index considering both the resource consumption and estimation precision.

References

1. J.M. Rabaey, M.J. Ammer, J.L. da Silva, D. Patel, S. Roundy, Picoradio supports ad hoc ultra-low power wireless networking. IEEE Comput **33**(7), 42–48 (2000)
2. A. Castagnetti, A. Pegatoquet, T.L. Le, M. Auguin, A joint duty-cycle and transmission power management for energy harvesting WSN. IEEE Trans. Ind. Inf. **10**(2), 928–936 (2014)
3. M. Magno, D. Boyle, D. Brunelli, E. Popovici, L. Benini, Ensuring survivability of resource-intensive sensor networks through ultra-low power overlays. IEEE Trans. Ind. Inf. **10**(2), 946–956 (2014)
4. W. Ikram, S. Petersen, P. Orten, N.F. Thornhill, Adaptive multi-channel transmission power control for industrial wireless instrumentation. IEEE Trans. Ind. Inf. **10**(2), 978–990 (2014)
5. J. Neuzil, O. Kreibich, R. Smid, A distributed fault detection system based on IWSN for machine condition monitoring. IEEE Trans. Ind. Inf. **10**(2), 1118–1123 (2014)
6. D. Ding, Z. Wang, B. Shen, Recent advances on distributed filtering for stochastic systems over sensor networks. Int. J. General Syst. **43**(3), 372–386 (2014)
7. I. Matei, J. Baras, Consensus-based linear distributed filtering. Automatica **48**(8), 1776–1782 (2012)
8. B. Chen, W. Zhang, L. Yu, Distributed finite-horizon fusion Kalman filtering for bandwidth and energy constrained wireless sensor networks. IEEE Trans. Signal Process. **62**(4), 797–812 (2014)
9. H. Dong, Z. Wang, H. Gao, Distributed filtering for a class of time-varying systems over sensor networks with quantization errors and successive packet dropouts. IEEE Trans. Signal Process **60**(6), 3164–3173 (2012)
10. W. Zhang, H. Dong, G. Guo, L. Yu, Distributed Sampled-data H_∞ filtering for sensor networks with nonuniform sampling periods. IEEE Trans. Ind. Inf. **10**(2), 871–881 (2014)
11. B. Shen, Z. Wang, Y.S. Hung, Distributed H_∞-consensus filtering in sensor networks with multiple missing measurements: the finite-horizon case. Automatica **46**(10), 1682–1688 (2010)
12. B. Kim, Y. Cho, J. Hong, AWNIS: Energy-efficient adaptive wireless network interface selection for industrial mobile devices. IEEE Trans. Ind. Inf. **10**(1), 714–729 (2013)
13. M. Tahir and S. Maxumder, Self-triggered communication enabled control of distributed generation in microgrids, IEEE Trans. Ind. Inf. (2015)
14. J. Li, G. AlRegib, Rate-constrained distributed estimation in wireless sensor networks. IEEE Trans. Signal Process. **55**(5), 1634–1643 (2007)
15. Q. Liu, Z. Wang, X. He, D.H. Zhou, Event-based recursive distributed filtering over wireless sensor networks. IEEE Trans. Autom. Control (2015)
16. M. Miskowicz, Send-on-delta concept: an event-based data reporting strategy. Sensors **6**(1), 49–63 (2006)
17. V.H. Nguyen, Y.S. Suh, Improving estimation performance in networked control systems applying the send-on-delta transmission method. Sensors **7**(10), 2128–2138 (2007)
18. Y.S. Suh, V.H. Nguyen, Y.S. Ro, Modified Kalman filter for networked monitoring systems employing a send-on-delta method. Automatica **43**(2), 332–338 (2007)
19. J. Sijs, M. Lazar, Event based state estimation with time synchronous updates. IEEE Trans. Autom. Control **57**(10), 2650–2655 (2012)
20. D. Yue, E. Tian, Q. Han, A delay system method for designing event-triggered controllers of networked control systems. IEEE Trans. Autom. Control **58**(2), 475–481 (2013)
21. D. Ding, Z. Wang, B. Shen, H. Dong, Envelope-constrained H_∞ filtering with fading measurements and randomly occurring nonlinearities: the finite horizon case. Automatica **55**, 37–45 (2015)
22. J. Wu, Q. Jia, K.H. Johansson, L. Shi, Event-based sensor data scheduling: trade-off between communication rate and estimation quality. IEEE Trans. Autom. Control **58**(4), 1041–1046 (2013)
23. W. Yu, G. Chen, M. Cao, J. Kurths, Second-order consensus for multi-agent systems with directed topologies and nonlinear dynamics. IEEE Trans. Syst. Man Cybern. Part B **40**(3), 881–891 (2010)

24. B.H. Kim, R. Baldick, A comparison of distributed optimal power flow algorithms. IEEE Trans. Power Syst. **15**(2), 599–604 (2000)
25. C. Drane, M. Macnaughtan, C. Scott, Positioning GSM telephones. IEEE Commun. Mag. **36**(4), 46–54 (1998)
26. A. De Luca, G. Oriolo, M. Vendittelli, *Control of wheeled mobile robots: An experimental overview* (Ramsete, Springer, Berlin Heidelberg, 2001), pp. 181–226

Chapter 6
Event-Based Distributed Filtering over Markovian Switching Topologies

It has now been well recognized that the communication networks may suffer from some degree of topological changes due to potentially harsh, uncertain, and dynamic wireless environments, and this gives rise to the so-called dynamic topology problem that has drawn considerable research interest in the context of consensus problems for multi-agent systems, see e.g., [1–4]. In [2], the stochastic link failure, which stems from the irregular detection areas of agents during the information exchange, has been thoroughly investigated. The dynamically changing interaction topologies has been addressed in [1], where the consensus has been shown to be achieved if the union of the directed interaction graphs has a spanning tree. As for the stochastic topologies, it has been proved in [3] that the mean topology determines whether the consensus of single-integrator agent can be achieved, and this result has then been extended to the case of Markovian switching topology in [4]. Although the issue of stochastically switching topologies has been attracting more attention in the community of multi-agent systems, the corresponding results concerning distributed filtering problems are scattered despite the profound application insights of wireless sensor networks.

Recently, some results have been available in the literature on the event-based distributed filtering problems, see, e.g., [5, 6]. Nonetheless, under the Markovian switching topologies, the topologies may stochastically change among a set of disconnected graphs at every sampling instant, which substantially impedes the information exchange between sensors, and thus hinders the cooperation of the distributed filters. Consequently, in order to guarantee the satisfactory filtering performance (e.g., mean-square boundedness of the estimation errors) for the distributed filtering dynamics, there is an urgent need to develop algorithms capable of accommodating the topology switching while achieving adequate estimation accuracy. As such, the focus of this chapter is on the event-based distributed filtering problems for wireless sensor networks over Markovian switching topologies.

Q. Liu et al., *Stochastic Control and Filtering over Constrained Communication Networks*, Studies in Systems, Decision and Control 178,
https://doi.org/10.1007/978-3-030-00157-5_6

The main challenges we are facing can be highlighted as follows: (1) how to define the triggering rule for each state with an individual threshold?; (2) what is the influence of the Markovian switching topologies on the distributed filters?; and (3) under which conditions the exponential mean-square boundedness of the estimation error dynamics for the distributed filtering system can be guaranteed? These challenges are properly handled in this chapter by developing a novel event-based distributed filtering algorithm. Specifically, a set of distributed event schedulers are deployed in every sensor to check its individual state so that the communication is only executed whenever an individual triggering condition is satisfied. By exploiting the stochastic analysis techniques and the graph theory, we conclude that the event-based distributed filters with Markovian switching topologies can achieve exponential mean-square boundedness if the sensor network is distributively detectable and the combined interaction topology is strongly connected. A numerical example is utilized to demonstrate the practical significance of the developed algorithm.

6.1 Problem Formulation

6.1.1 Markovian Switching Topologies

Directed graphs can be utilized to describe the communication connections of the individual sensors in sensor networks. A directed graph is denoted by $\mathcal{G} = (\mathcal{V}, \mathcal{E}, \mathcal{A})$ with a vertex set $\mathcal{V} = \{v_1, v_2, \ldots, v_N\}$, an edge set $\mathcal{E} \subseteq \mathcal{V} \times \mathcal{V}$, and an adjacency matrix $\mathcal{A}= [a_{ij}]_{N\times N}$ with nonnegative elements a_{ij}. The edge set represents the topology connections between nodes. Note that $(v_i, v_j) \in \mathcal{E}$, if and only if $a_{ij} > 0$, indicating that the ith node can receive the data from the jth node. We do not allow the self-loops, namely, $a_{ii} = 0, \forall i \in \mathcal{V}$. The set of neighbors of ith node is denoted by $N_i = \{v_j \in \mathcal{V} | (v_i, v_j) \in \mathcal{E}\}$. The Laplacian matrix for $\mathcal{G}$ is defined as $H = \mathcal{D} - \mathcal{A}$, where $\mathcal{D} = \mathrm{diag}\{d_1, d_2 \ldots, d_N\}$ is a diagonal matrix with the entries $d_i = \sum_{j=1}^{N} a_{ij}$. The graph is said to be strongly connected if every node is reachable from every other node.

In the chapter, we suppose that the sensor nodes are deployed with the communication connections subject to a set of time-varying graphs $\mathcal{G}(r(t)) = (\mathcal{V}, \mathcal{E}(r(t)), \mathcal{A}(r(t)))$. The evolution of the graph $\mathcal{G}(r(t))$ is governed by a homogeneous continuous-time Markov process $\{r(t)\}$ (with right continuous trajectories) taking values on a finite set $\mathbb{S} = \{1, 2, \ldots, S\}$. As such, we have the corresponding adjacency matrix $\mathcal{A}(r(t))$, the set of neighbors $N_i(r(t))$ and the Laplacian matrix $H(r(t))$. The transition probabilities of the Markov process are given by

$$\mathbb{P}\{r(t + \Delta t) = j | r(t) = i\} = \begin{cases} \pi_{ij}\Delta t + o(\Delta t), & \text{if } i \neq j \\ 1 + \pi_{ii}\Delta t + o(\Delta t), & \text{elsewise} \end{cases}$$

where $[\pi_{ij}]_{S\times S}$ is a finite-dimensional transition rate matrix of the Markov chain with $\pi_{ij} \geq 0$ for $i \neq j$ and $\pi_{ii} = -\sum_{j\neq i} \pi_{ij}$. The term $o(\Delta t)$ represents an infinitesimal of higher order than Δt, i.e., $\lim_{\Delta t\to 0} o(\Delta t)/\Delta t = 0$.

6.1.2 Wireless Sensor Networks

Consider the target plant described by the following nonlinear continuous-time stochastic system:

$$\dot{x}(t) = f(x(t), t) + b(t)w(t), \tag{6.1}$$

where $x(t) = (x_1(t), x_2(t), \ldots, x_n(t))' \in \mathbb{R}^n$ is the internal state of the plant, $f(x) = (f_1(x), f_2(x), \ldots, f_n(x))' \in \mathbb{R}^n$ is a nonlinear function which contributes to the dynamical evolution of the target plant, $b(t) = (b_1(t), b_2(t), \ldots, b_n(t))' \in \mathbb{R}^n$ represents the noise intensity, and $w(t)$ is a scalar Gaussian random variable with zero mean and autocorrelation $\mathbb{E}\{w(t)w(t+\tau)\} = \delta(\tau)$, where $\delta(\cdot)$ is a Dirac delta function with the properties

$$\delta(t) = \begin{cases} 0 & \text{for } t \neq 0 \\ \infty & \text{for } t = 0 \end{cases} \quad \text{and} \quad \int_{-\infty}^{\infty} \delta(t)dt = 1.$$

For every $i \in \mathcal{V}$, the measurement of sensor i is given as follows:

$$y_i(t) = C_i x(t) + v_i(t)\zeta_i(t), \tag{6.2}$$

where $y_i(t) \in \mathbb{R}^{m_i}$ is the measurement signal, $v_i(t) = (v_{i1}(t), v_{i2}(t), \ldots, v_{in}(t))' \in \mathbb{R}^n$ represents the noise intensity, and $\zeta_i(t) \in \mathbb{R}$ is a scalar Gaussian random variable with zero mean and autocorrelation $\mathbb{E}\{\zeta_i(t)\zeta_j(t+\tau)\} = \delta(i-j)\delta(\tau)$. C_i is the measurement matrix of appropriate dimensions. The random variables $r(t)$, $w(t)$, and $\zeta_i(t)$ are mutually independent, and the components of the vectors $b(t)$ and $v_i(t)$ are bounded, i.e., $0 < b_k(t) \leq \bar{b}_k$ and $0 < v_{ik}(t) \leq \bar{v}_{ik}$, for $i \in \mathcal{V}$ and $k = 1, 2, \ldots, n$, where $\bar{b}_k$ and $\bar{v}_{ik}$ are positive scalars. The system (6.1)–(6.2) under consideration is assumed to satisfy the following assumptions.

Assumption 6.1 ([7]) The nonlinear function $f(\cdot) : \mathbb{R}^n \to \mathbb{R}^n$ is analytic everywhere and satisfies the following condition:

$$\begin{aligned} &(x-y)'P(f(x,t) - f(y,t) - \Delta(x-y)) \\ &\qquad \leq -\eta(x-y)'(x-y), \quad \forall x, y \in \mathbb{R}^n, \end{aligned} \tag{6.3}$$

where $P = \text{diag}\{p_1, p_2, \ldots, p_n\}$, $\Delta = \text{diag}\{\delta_1, \delta_2, \ldots, \delta_n\}$, and η, p_i, δ_i are positive scalars.

Remark 6.1 It should be stressed that Assumption 6.1 is very mild. If $\partial f_i/\partial x_i$, $i = 1, 2, \ldots, n$ are bounded, then this assumption must be satisfied.

Assumption 6.2 The continuous-time Markov process $\{r(t)\}$ with the transition rate matrix $[\pi_{ij}]_{S\times S}$ is ergodic.

Remark 6.2 Under the ergodic assumption, the state of the Markov process can be reached from any other state and, for any initial distribution, there always exists a unique stationary distribution $\{\bar{\pi}_i > 0, i \in \mathbb{S}\}$ with $\sum_{i=1}^{n} \bar{\pi}_i = 1$. Moreover, when the Markov process is started off initially with such a stationary distribution, the distribution will be invariant over time.

6.1.3 Event-Based Distributed Filter

A fundamental issue of the filtering problems is to estimate the state vector $x(t)$ based on the noisy measurements. It is noted that different from the classical single filter case, the target plant is now observed by a group of smart sensors, and the estimation is carried out in every sensor node in a distributive fashion. Due to the physical limitations of the sensors, the individual sensor usually has insufficient abilities to estimate the plant state based on the local measurements only. To achieve high accuracy estimation, we need to utilize the complementary messages gathered from all the local and neighboring nodes. As such, the distributed filtering algorithm is chosen as follows:

$$\begin{cases} \dfrac{d\hat{x}_i(t)}{dt} = f(\hat{x}_i(t), t) + cL_i(y_i(t) - C_i\hat{x}_i(t)) \\ \qquad\qquad + c \displaystyle\sum_{j\in N_i(r(t))} a_{ij}(r(t))(\hat{x}_j(t) - \hat{x}_i(t)) \\ \hat{x}_i(0) = 0 \end{cases} \tag{6.4}$$

where $\hat{x}_i(t) \in \mathbb{R}^n$ is the local estimate of the full state vector $x(t)$ from the ith sensor. The filter gain $L_i \in \mathbb{R}^{n\times m}$ and the strength $c \in \mathbb{R}$ are parameters to be designed. According to the above structure, every sensor calculates the local estimate based on its measurements $y_i(t)$ as well as the estimates $\hat{x}_i(t)$ received from the neighboring sensors.

When taking the energy and bandwidth restrictions into account, some new challenges arise for the distributed filtering problem. As shown in (6.4), in order to compute the local state estimates, one needs to use the real-time neighboring estimates transmitted via wireless networks. Note that the network communication constitutes one of the main sources for the consumption of the constrained resources. Too frequent transmissions will inevitably put more burden on the network and deteriorate the network performance leading to some unfavorable phenomena such as packet losses and communication delays. As such, in this chapter, an event-based scheme is introduced to determine whether the local information will be broadcast or not.

To begin with, we define a sequence of event instants for the sth entry of the estimate $\hat{x}_i(t)$ (i.e., $\hat{x}_{i,s}(t)$) by a monotonically increasing sequence $0 = t_{i,s}^0 < t_{i,s}^1 < t_{i,s}^2 < \cdots$ and the broadcast is triggered only at these event instants. Such a sequence is determined iteratively online according to the following rule $t_{i,s}^{k+1} = \inf\{t \in \mathbb{R}^+ | t > t_{i,s}^k, \hbar_{i,s}(\hat{x}_{i,s}(t), \hat{x}_{i,s}(t_{i,s}^k), t) > 0\}$, where the event function $\hbar_{i,s}(\cdot) : \mathbb{R} \times \mathbb{R} \times \mathbb{R}^+ \to \mathbb{R}$ is chosen to be

$$\hbar_{i,s}(\hat{x}_{i,s}(t), \hat{x}_{i,s}(t_{i,s}^k), t) = (\hat{x}_{i,s}(t) - \hat{x}_{i,s}(t_{i,s}^k))^2 - \sigma_{i,s}$$

for $i \in \mathcal{V}$ and $s = 1, 2, \ldots, n$ with the individual thresholds $\sigma_{i,s}$ being positive scalars. From the above event-based mechanism, it can be seen that, whenever the triggering condition $\hbar_{i,s}(\hat{x}_{i,s}(t), \hat{x}_{i,s}(t_{i,s}^k), t) > 0$ is satisfied, a broadcast of $\hat{x}_{i,s}(t)$ is triggered with the marked event instant $t_{i,s}^{k+1}$ such that $\hat{x}_{i,s}(t) = \hat{x}_{i,s}(t_{i,s}^{k+1})$, and thus that $\hbar_{i,s}(\hat{x}_{i,s}(t), \hat{x}_{i,s}(t_{i,s}^{k+1}), t)$ is equal to $-\sigma_{i,s}$. As such, the following inequalities are always true

$$\hbar_{i,s}(\hat{x}_{i,s}(t), \hat{x}_{i,s}(t_{i,s}^k), t) \leq 0.$$

For the purpose of clarity, we augment all the components at the latest event instants $\hat{x}_i^t(t) \triangleq \mathrm{col}\{\hat{x}_{i,1}(t_{i,1}^{k_{i,1}(t)}), \ldots, \hat{x}_{i,n}(t_{i,n}^{k_{i,n}(t)})\}$. where $k_{i,j}(t) = \max\{\kappa \in \mathbb{N} \mid t_{i,j}^{\kappa} < t\}$. According to the event-based mechanism, let us revise (6.4) by only using the neighboring information at their latest event instants. In the event-based case, the distributed filter is given as follows:

$$\begin{aligned}\frac{d\hat{x}_i(t)}{dt} &= f(\hat{x}_i(t),t) + cL_i(y_i(t) - C_i\hat{x}_i(t)) \\ &\quad + c\sum_{j\in N_i(r(t))} a_{ij}(r(t))(\hat{x}_j^t(t) - \hat{x}_i^t(t)),\end{aligned} \tag{6.5}$$

which is further rewritten as below:

$$\begin{aligned}\frac{d\hat{x}_i(t)}{dt} =& f(\hat{x}_i(t), t) - c\sum_{j=1}^{N} h_{ij}(r(t))\hat{x}_j^t(t) \\ &+ cL_i(y_i(t) - C_i\hat{x}_i(t)),\end{aligned} \tag{6.6}$$

where $h_{ij}(r(t))$ is the (i, j)th element of the Laplacian matrix $H(r(t))$ associated with the graph $\mathcal{G}(r(t))$.

Denote the event-induced and estimation errors by $e_i^t(t) \triangleq \hat{x}_i^t(t) - \hat{x}_i(t)$ and $e_i(t) \triangleq \hat{x}_i(t) - x(t)$, respectively. By subtracting (6.6) from (6.1), we can obtain the dynamics of the estimation error as follows:

$$\begin{aligned}\frac{de_i(t)}{dt} &= f(\hat{x}_i(t), t) - f(x_i(t), t) - cL_iC_ie_i(t) \\ &\quad - c\sum_{j=1}^{N} h_{ij}(r(t))e_j(t) + cL_iv_i(t)\zeta_i(t) \\ &\quad - c\sum_{j=1}^{N} h_{ij}(r(t))e_j^t(t) - b(t)w(t). \end{aligned} \tag{6.7}$$

Before proceeding, we introduce the following definition for the exponential mean-square boundedness.

Definition 6.1.1 The dynamics of the estimation error in (6.7) is said to be *exponentially bounded in mean square* if there exist real numbers $\rho > 0$, $\tau > 0$ and $\mu > 0$ such that

$$\sum_{i=1}^{N} \mathbb{E}\{\|e_i(t)\|^2\} \leq \rho e^{-\tau t} \sum_{i=1}^{N} \mathbb{E}\{\|e_i(0)\|^2\} + \mu. \tag{6.8}$$

where τ is the decay rate and μ is the ultimate bound.

The objective of this chapter is to design the parameter c and the filter gains L_i for each sensor such that the dynamics of the distributed estimation error (6.7) is exponentially bounded in the mean-square sense.

6.2 Main Results

In this section, a sufficient criterion will be established to guarantee the exponential mean-square boundedness of the estimation errors for the proposed event-based distributed filtering algorithm over Markovian switching topologies.

A useful lemma is presented as follows.

Lemma 6.2.1 ([8]) *Suppose that a directed graph $\mathcal{G}$ with the Laplacian matrix H is strongly connected. Then, there exists a positive vector $\xi = (\xi_1, \xi_2, \ldots, \xi_N)'$ such that $\xi' H = 0$. Furthermore, $(H)^s \triangleq (\Xi H + H'\Xi)$ is a symmetric matrix with zero row sum, where $\Xi = \mathrm{diag}\{\xi_1, \xi_2, \ldots, \xi_N\}$.*

To continue the discussion, we need to introduce the definition for the distributive detectability as follows:

Definition 6.2.2 A sensor network described by (6.1)–(6.2) is said to be *distributively detectable* if there exist matrices L_i satisfying

$$
\begin{aligned}
&PL_iC_i + C_i'L_i'P \geq 2 \sum_{k \in \mathcal{M}_i} \Theta_{ik}, \\
&\bigcup_{\forall i \in \mathcal{V}} \mathcal{M}_i = \{k\}_{k=1}^{n},
\end{aligned}
\tag{6.9}
$$

where P is a positive definite diagonal matrix defined in Assumption 6.1, $\mathcal{M}_i$ is any arbitrary subset of $\{k\}_{k=1}^{n}$, and Θ_{ik} is a diagonal matrix of the form

$$
\Theta_{ik} \triangleq \mathrm{diag}\{\underbrace{0, \ldots, 0}_{k-1}, \gamma_{ik}, \underbrace{0, \ldots, 0}_{n-k}\},
$$

with a positive real number $\gamma_{ik} > 0$.

The distributive detectability property plays an essential role in guaranteeing the boundedness of the estimation error dynamics. For a distributive detectable system, the following condition always holds:

$$
x'(t)P\Big(L_iC_ix(t)\Big) \geq \sum_{k \in \mathcal{M}_i} \gamma_{ik} x_k'(t) x_k(t), \tag{6.10}
$$

from which we can see that the kth ($k \in \mathcal{M}_i$) entry of the state vector can be extracted by sensor i. Notice that the union of all the subsets equals to the whole set, i.e., $\bigcup_{\forall i \in \mathcal{V}} \mathcal{M}_i = \{k\}_{k=1}^{n}$. In this case, we will show that the sensor network can complete the distributed estimation from a cooperative perspective.

For simplicity, we introduce the notations

$$
\begin{aligned}
e(t) &\triangleq \mathrm{col}\{e_1(t), \ldots, e_N(t)\} \\
e^t(t) &\triangleq \mathrm{col}\{e_1^t(t), \ldots, e_N^t(t)\} \\
\tilde{e}_k(t) &\triangleq \mathrm{col}\{e_{1,k}(t), \ldots, e_{N,k}(t)\} \\
\widehat{\Theta}_{ik} &\triangleq \mathrm{diag}\{\underbrace{0, \ldots, 0}_{i-1}, \gamma_{ik}, \underbrace{0, \ldots, 0}_{N-i}\},
\end{aligned}
$$

where $e_{i,k}(t)$ is the kth element of the vector $e_i(t)$. Additionally, we denote the set gathering all the sensors that can extract the kth entry of the target state by $\mathcal{M}_k^- \triangleq \{i\ : k \in \mathcal{M}_i,\ \text{for } i \in \mathcal{V}\}$.

When the sensor network has the distributive detectability as defined in Definition 6.2.2, it can be verified that $\bigcup_{\forall k \in \{1,\ldots,n\}} \mathcal{M}_k^- = \mathcal{V}$. Furthermore, the ergodicity of the Markov chain ensures that there exists a unique stationary distribution $\{\bar{\pi}_i > 0,\ i \in \mathbb{S}\}$. Without loss of generality, we assume that the Markov process starts from such a stationary distribution. As such, it can be seen that $\mathbb{P}\{r(t) = i\} = \bar{\pi}_i > 0$, $\forall t \geq 0$. Moreover, we define the weighted Laplacian matrix as $\widehat{H} \triangleq \sum_{l=1}^{S} \bar{\pi}_l H(l)$ whose corresponding graph is called *the combined graph*. According to Lemma 6.2.1, we can compute the matrix Ξ with respect to the weighted matrix $\widehat{H}$ if the combined graph is strongly connected. Let $\xi^- = \min\{\xi_1, \xi_2, \ldots, \xi_n\}$, $\xi^+ = \max\{\xi_1, \xi_2, \ldots, \xi_n\}$, $p^- = \min\{p_1, p_2, \ldots, p_n\}$, and $p^+ = \max\{p_1, p_2, \ldots, p_n\}$.

Now, we are in the position to obtain a sufficient condition for the exponential mean-square boundedness in the following theorem.

Theorem 6.2.3 *Consider the system* (6.1)–(6.2) *subject to Markovian switching topologies. Under Assumptions 6.1–6.2, if the following criteria:*

$$(2p_k\delta_k + 1)\Xi - c\left(p_k\widehat{H} + \sum_{i\in\mathcal{M}_k^-}\widehat{\Theta}_{ik}\right)^s < 0 \tag{6.11}$$

are satisfied for $k = 1, 2, \ldots, n$, then the filtering error dynamics (6.7) is exponentially mean-square bounded with the decay rate $\tau < 2\eta/p^+$ and the ultimate bound $\mu = \tilde{\mu}/\tau\xi^- p^-$ where

$$\begin{aligned}\tilde{\mu} =& \sum_{i=1}^{N}\sum_{k=1}^{n}\xi_i p_k \bar{b}_k^2 + c^2 p^+ \sum_{i=1}^{N}\sum_{k=1}^{n}\xi_i \lambda_{max}(L_i' L_i)\bar{v}_{ik}^2 \\ &+ c^2\xi^+\lambda_{\max}(\widehat{H}'\widehat{H})\sum_{i=1}^{N}\sum_{k=1}^{n} p_k\sigma_{i,k}.\end{aligned} \tag{6.12}$$

Proof Construct the following Lyapunov-like function for system (6.7):

$$V(e(t), j) = \sum_{i=1}^{N}\xi_i e_i'(t)P_j e_i(t). \tag{6.13}$$

where P_j are positive definite matrices. Denote the infinitesimal operator $\mathcal{L}(\cdot)$ of the random process $\{e(t), r(t)\}$ by

$$\begin{aligned}&\mathcal{L}V(e(t), j)\\ &\triangleq \lim_{\Delta t\to 0^+}\frac{1}{\Delta t}\Big\{\mathbb{E}\{V(e(t+\Delta t), r(t+\Delta t))|r(t) = j, e(t)\}\\ &\quad - V(e(t), j)\Big\},\end{aligned}$$

then, we have (see [9, 10])

$$
\begin{aligned}
&\mathcal{L}V(e(t), j)\\
&= \sum_{i=1}^{N} \Big\{ 2\xi_i e_i'(t) P_j \big(f(\hat{x}_i(t), t) - f(x_i(t), t)\\
&\quad - c\sum_{j=1}^{N} h_{ij}(r(t)) e_j(t) - c\sum_{j=1}^{N} h_{ij}(r(t)) e_j^t(t)\\
&\quad - cL_i C_i e_i(t)\big) + c^2 \xi_i v_i'(t) L_i' P_j L_i v_i(t)\\
&\quad + \xi_i b'(t) P_j b(t) \Big\} + \sum_{l=1}^{S} \pi_{jl} V(e(t), l).
\end{aligned} \tag{6.14}
$$

Letting $P_1 = P_2 = \cdots = P$ and, from Assumption 6.1, we have

$$
\begin{aligned}
&e_i'(t) P \big(f(\hat{x}_i(t), t) - f(x_i(t), t) \big)\\
&\quad \le -\eta e_i'(t) e_i(t) + e_i'(t) P \Delta e_i(t).
\end{aligned} \tag{6.15}
$$

Moreover, through some algebraic manipulations, it can be verified that

$$
\begin{aligned}
&2\sum_{i=1}^{N} \xi_i e_i'(t) P \Delta e_i(t)\\
&= 2\sum_{i=1}^{N} \sum_{k=1}^{n} \xi_i e_{i,k}'(t) p_k \delta_k e_{i,k}(t)\\
&= 2\sum_{k=1}^{n} p_k \delta_k \tilde{e}_k'(t) \Xi \tilde{e}_k(t),
\end{aligned} \tag{6.16}
$$

and

$$
\begin{aligned}
&2\sum_{i=1}^{N} \xi_i e_i'(t) P \Big(c\sum_{j=1}^{N} h_{ij}(r(t)) e_j(t) \Big)\\
&= 2c \sum_{i=1}^{N} \sum_{j=1}^{N} \sum_{k=1}^{n} \xi_i e_{i,k}'(t) p_k h_{ij}(r(t)) e_{j,k}(t)\\
&= c\sum_{k=1}^{n} p_k \tilde{e}_k'(t) (H(r(t)))^s \tilde{e}_k(t),
\end{aligned} \tag{6.17}
$$

where $(H(r(t)))^s = \Xi H(r(t)) + H'(r(t))\Xi$. Furthermore, we have

$$\begin{aligned}\sum_{i=1}^{N}\xi_i b'(t)Pb(t) &= \sum_{i=1}^{N}\sum_{k=1}^{n} p_k\xi_i b_k'(t)b_k(t)\\ &\le \sum_{i=1}^{N}\sum_{k=1}^{n} p_k\xi_i \bar{b}_k^2 \end{aligned}\tag{6.18}$$

and

$$\begin{aligned}&\sum_{i=1}^{N} c^2\xi_i v_i'(t)L_i'PL_iv_i(t)\\ &\le c^2p^+\sum_{i=1}^{N}\sum_{k=1}^{n}\xi_i\lambda_{max}(L_i'L_i)v_{ik}'(t)v_{ik}(t)\\ &\le c^2p^+\sum_{i=1}^{N}\sum_{k=1}^{n}\xi_i\lambda_{max}(L_i'L_i)\bar{v}_{ik}^2. \end{aligned}\tag{6.19}$$

Note that the Markov process $\{r(t)\}$ starts from the invariant distribution $\{\bar{\pi}_i > 0,\ i \in \mathbb{S}\}$. According to [9] and [11], we have

$$\mathbb{E}\{\mathcal{L}V(e(t), r(t))\} = \sum_{j=1}^{S}\mathbb{E}\left\{\mathcal{L}V(e(t), j)\bar{\pi}_j\right\}.$$

Substituting (6.14)–(6.19) into the above equations and together with $\sum_{j=1}^{S}\pi_{ij} = 0$, we obtain

$$\begin{aligned}&\mathbb{E}\{\mathcal{L}V(e(t), r(t))\}\\ &\le \mathbb{E}\Big\{-2\eta\sum_{i=1}^{N}\xi_i e_i'(t)e_i(t)\Big\} - \mathbb{E}\Big\{2c\sum_{i=1}^{N}\xi_i e_i'(t)PL_iC_ie_i(t)\Big\}\\ &+\mathbb{E}\Big\{2\sum_{k=1}^{n}p_k\tilde{e}_k'(t)\Big(2\delta_k\Xi - c(\widehat{H})^s\Big)\tilde{e}_k(t)\Big\}\\ &+\mathbb{E}\Big\{2\sum_{i=1}^{N}\xi_i e_i'(t)P\Big(c\sum_{j=1}^{N}\hat{h}_{ij}e_j'(t)\Big)\Big\}\\ &+\sum_{i=1}^{N}\sum_{k=1}^{n}c^2p^+\xi_i\lambda_{max}(L_i'L_i)\bar{v}_{ik}^2 + \sum_{i=1}^{N}\sum_{k=1}^{n}p_k\xi_i\bar{b}_k^2, \end{aligned}\tag{6.20}$$

where $\hat{h}_{ij}$ is the (i, j)th element of the matrix $\widehat{H}$. Invoking the basic inequality $(x + y)'(x + y) \ge 0$, one derives that

$$-2\sum_{i=1}^{N}\xi_i e_i'(t)P\Big(c\sum_{j=1}^{N}\hat{h}_{ij}e_j^t(t)\Big)\leq\sum_{i=1}^{N}\xi_i e_i'(t)Pe_i(t)$$
$$+\sum_{i=1}^{N}\xi_i\Big(c\sum_{j=1}^{N}\hat{h}_{ij}e_j^t(t)\Big)'P\Big(c\sum_{j=1}^{N}\hat{h}_{ij}e_j^t(t)\Big). \tag{6.21}$$

The second term in the right-hand side of the above inequality can be rewritten as follows:

$$\begin{aligned}
&\sum_{i=1}^{N}\xi_i\Big(c\sum_{j=1}^{N}\hat{h}_{ij}e_j^t(t)\Big)'P\Big(c\sum_{j=1}^{N}\hat{h}_{ij}e_j^t(t)\Big)\\
&\quad=c^2e^t(t)'\Big((\widehat{H}'\Xi\widehat{H})\otimes P\Big)e^t(t)\\
&\quad\leq c^2\xi^+\lambda_{\max}(\widehat{H}'\widehat{H})\sum_{i=1}^{N}\sum_{k=1}^{n}p_k\sigma_{i,k},
\end{aligned} \tag{6.22}$$

where the last inequality follows from $\hbar_{i,s}(\hat{x}_{i,s}(t),\hat{x}_{i,s}(t_k^{(i,s)}),t)\leq 0$. Substituting (6.21)–(6.22) into (6.20) yields that

$$\begin{aligned}
&\mathbb{E}\{\mathcal{L}V(e(t),r(t))\}\\
&\leq\mathbb{E}\Big\{\sum_{k=1}^{n}p_k\tilde{e}_k'(t)\left(2\delta_k\Xi-c(\widehat{H})^s+\Xi\right)\tilde{e}_k(t)\Big\}\\
&-\mathbb{E}\Big\{2c\sum_{i=1}^{N}\xi_i e_i'(t)PL_iC_ie_i(t)\Big\}+\tilde{\mu}\\
&-\mathbb{E}\Big\{2\eta\sum_{i=1}^{N}\xi_i e_i'(t)e_i(t)\Big\},
\end{aligned} \tag{6.23}$$

By utilizing the distributed detectability of the sensor network (6.9), we obtain

$$\begin{aligned}
&-2\sum_{i=1}^{N}\xi_i e_i'(t)P\Big(cL_iC_ie_i(t)\Big)\\
&\leq-2c\sum_{i=1}^{N}\sum_{k\in\mathcal{M}_i}\gamma_{ik}\xi_i e_{i,k}'(t)e_{i,k}(t)\\
&=-2c\sum_{k=1}^{n}\sum_{i\in\mathcal{M}_k^-}\gamma_{ik}\xi_i e_{i,k}'(t)e_{i,k}(t)
\end{aligned}$$

$$= -2c\sum_{k=1}^{n}\tilde{e}'_k(t)\Big(\Xi\sum_{i\in\mathcal{M}_k^-}\widehat{\Theta}_{ik}\Big)\tilde{e}_k(t). \tag{6.24}$$

As such, when the condition in (6.11) is fulfilled, the following inequality can be derived

$$\begin{aligned}&\mathbb{E}\left\{\mathcal{L}V(e(t),r(t))\right\}\\&\leq -2\eta\mathbb{E}\Big\{\sum_{i=1}^{N}\xi_i e'_i(t)e_i(t)\Big\}+\tilde{\mu}.\end{aligned} \tag{6.25}$$

To further prove the exponential mean-square boundedness, we construct a new function as follows:

$$\mathscr{V}(t,e(t),r(t)) \triangleq e^{\tau t}V(e(t),r(t)),$$

where $\tau > 0$ is a constant to be determined later. It is straightforward to show that

$$\begin{aligned}&\mathbb{E}\{\mathcal{L}\mathscr{V}(t,e(t),r(t))\}\\&\quad= e^{\tau t}\mathbb{E}\{\mathcal{L}V(e(t),r(t))\}+\tau e^{\tau t}\mathbb{E}\{V(e(t),r(t))\}.\end{aligned}$$

Note that

$$\mathbb{E}\{V(e(t),r(t))\}\leq p^{+}\mathbb{E}\Big\{\sum_{i=1}^{N}\xi_i e'_i(t)e_i(t)\Big\}.$$

Choosing $\tau < 2\eta/p^+$, in conjunction with (6.25), it is not difficult to prove that

$$\begin{aligned}&\mathbb{E}\{\mathcal{L}\mathscr{V}(t,e(t),r(t))\}\\&\leq(-2\eta+\tau p^{+})e^{\tau t}\mathbb{E}\Big\{\sum_{i=1}^{N}\xi_i e'_i(t)e_i(t)\Big\}+e^{\tau t}\tilde{\mu}\leq e^{\tau t}\tilde{\mu}.\end{aligned}$$

By using the generalized Itô formula [12], one has

$$\begin{aligned}&\mathbb{E}\left\{\mathscr{V}(t,e(t),r(t))\right\}\\&=\mathbb{E}\left\{\mathscr{V}(0,e(0),r(0))\right\}+\int_0^t\mathbb{E}\left\{\mathcal{L}\mathscr{V}(s,e(s),r(s))\right\}ds\\&\leq\mathbb{E}\left\{V(e(0),r(0))\right\}+\tilde{\mu}\int_0^t e^{\tau s}ds.\end{aligned} \tag{6.26}$$

Obviously, we have $\mathbb{E}\{V(e(t),\ r(t))\}\geq\xi^{-}p^{-}\sum_{i=1}^{N}\mathbb{E}\{\|e_i(t)\|^2\}$ and $\mathbb{E}\{V(e(0),$ $r(0))\}\leq\xi^{+}p^{+}\sum_{i=1}^{N}\mathbb{E}\{\|e_i(0)\|^2\}$. Thus, from (6.26), it follows that

$$\sum_{i=1}^{N} \|e_i(t)\|^2 \leq \frac{\xi^+ p^+}{\xi^- p^-} e^{-\tau t} \sum_{i=1}^{N} \|e_i(0)\|^2 + \mu,$$

which eventually implies that the distributed filtering system is exponentially mean-square bounded. The proof is complete.

The following corollary is readily accessible from Theorem 6.2.3.

Corollary 6.2.4 *Consider the event-based distributed filtering system over Markovian switching topologies. If the sensor network is distributively detectable and the union of the graphs $\{\mathcal{G}_1, \ldots, \mathcal{G}_S\}$ is strongly connected, then by choosing the parameter*

$$c > \max_k \frac{(2p_k\delta_k + 1)\xi^+}{\lambda_{\min}\left\{\left(p_k\widehat{H} + \sum_{i\in\mathcal{M}_k^-}\widehat{\Theta}_{ik}\right)^s\right\}} \tag{6.27}$$

the filtering error dynamics (6.7) is exponentially mean-square bounded.

Proof Because of the distributive detectability of the sensor networks, every component of the state can be observed by at least a sensor node, i.e., $\mathcal{M}_k^- \neq \emptyset$, for $k = 1, 2, \ldots, n$, where $\emptyset$ represents an empty set. Since the combined graph is strongly connected, the weighted Laplacian matrix $\widehat{H}$ is irreducible. According to Lemma 6.2.1, it is not hard to verify that the matrix $(p_k\widehat{H})^s$ is a new symmetric Laplacian matrix with the eigenvalues satisfying $\lambda_N((p_k\widehat{H})^s) \geq \cdots, \geq \lambda_2((p_k\widehat{H})^s) > \lambda_1((p_k\widehat{H})^s) = 0$. As a result, we have $\left(p_k\widehat{H} + \sum_{i\in\mathcal{M}_k^-}\widehat{\Theta}_{ik}\right)^s > 0$. To this end, it can be seen that if we choose c according to (6.27), then the criteria (6.11) is satisfied and therefore the dynamics of the estimation error is exponentially mean-square bounded, which ends the proof.

Remark 6.3 In the proposed distributed filter, there are two parameters (i.e., strength c and filter gains L_i) waited to be designed, where L_i can be obtained by solving the inequalities in (6.9) and c can be determined based on (6.27) in Corollary 6.2.4. That is to say, if the conditions in Corollary 6.2.4 are satisfied, then by choosing the parameter $c > \max_k (2p_k\delta_k + 1)\xi^+/\lambda_{\min}\left\{\left(p_k\widehat{H} + \sum_{i\in\mathcal{M}_k^-}\widehat{\Theta}_{ik}\right)^s\right\}$, the filtering error dynamics must be exponentially mean-square bounded.

Remark 6.4 Note that the threshold $\sigma_{i,s}$ determines the size of an event domain. A large $\sigma_{i,s}$ will result in the increase of ultimate bound $\tilde{\mu}$ (see (6.12)) while usually reduce the communication frequency. Particularly, when the thresholds $\sigma_{i,s} = 0$, the event-based transmission strategy reduces to the classical clock-driven one.

Remark 6.5 For estimator design problems of stochastic nonlinear systems subject to Brownian perturbations, an excellent work [13] has been carried out which provides a rigorous mathematical proof of the finite-time input-to-stabilization of the estimation error dynamics. Different from [13], we consider the distributed state estimation in this chapter where the nonlinear stochastic system under consideration is subject to both Brownian perturbations and Markovian switching topologies.

6.3 An Illustrative Example

In this section, a numerical example is provided to demonstrate the applicability of the proposed filtering technique.

The sensor network under consideration is composed of $N = 4$ nodes. The directed graphs $\mathcal{G}(r(t)) = (\mathcal{V}, \mathcal{E}(r(t)), \mathcal{A}(r(t)))$ with $r(t) \in \{1, 2\}$ are depicted in Fig. 6.1. The adjacency matrices $\mathcal{A}_1$ and $\mathcal{A}_2$ are given by

$$\mathcal{A}_1 = \begin{bmatrix} 0 & 0 & 0 & 0 \\ 1 & 0 & 0 & 0 \\ 0 & 0 & 0 & 1 \\ 0 & 1 & 0 & 0 \end{bmatrix}, \quad \mathcal{A}_2 = \begin{bmatrix} 0 & 0 & 1 & 1 \\ 0 & 0 & 0 & 0 \\ 0 & 0 & 0 & 1 \\ 0 & 0 & 0 & 0 \end{bmatrix}.$$

The transmission probability matrix is chosen to be

$$[\pi_{ij}]_{2\times 2} = \begin{bmatrix} -1 & 1 \\ 1 & -1 \end{bmatrix}.$$

Suppose that the initial distribution of the Markov process obeys an invariant distribution $\bar{\pi}_1 = 0.5$ and $\bar{\pi}_2 = 0.5$. The dynamics of the target system is

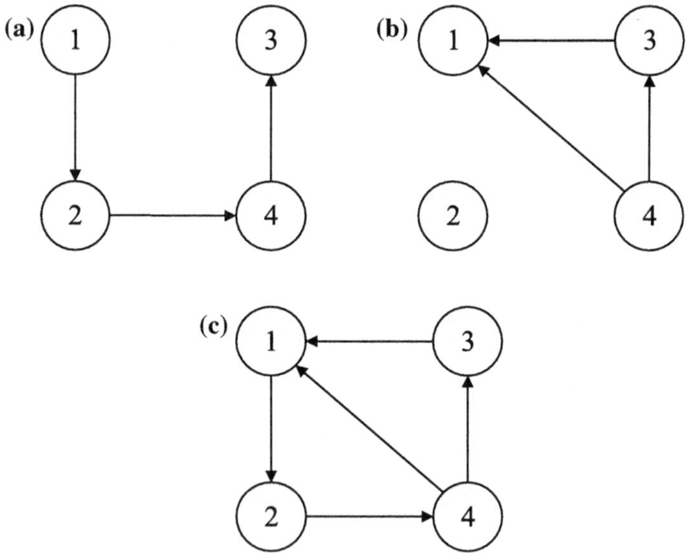

Fig. 6.1 **a** and **b** are the topology of $\mathcal{G}_1$ and $\mathcal{G}_2$, respectively. **c** is the topology of the combined graph

Table 6.1 Triggering times in total 2, 000 Steps

Sensor	1th	2th	3th	4th
First element	321	375	747	413
Second element	1130	705	1084	1286

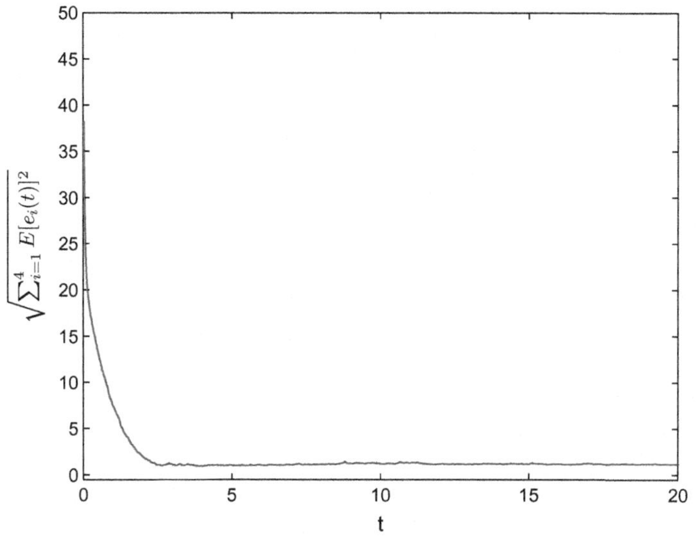

Fig. 6.2 The evolution of the mean-square error

$$\begin{cases} \dot{s} = -s + 0.2132\alpha + 0.1521s^2\alpha^2 + b_1 w, \\ \dot{\alpha} = -0.5000\alpha - 0.1018\alpha^3 + b_2 w, \end{cases}$$

where s represents the position and α is the velocity with the respective initial values -3.70 and 5.50. It is not hard to verify that the above dynamics satisfies the constraint in Assumption 6.1 with $P = \text{diag}\{1, 1\}$, $\Delta = \text{diag}\{8, 7\}$, and $\eta = 0.01$. The measurement matrices are chosen to be

$$C_1(k) = [10\ 0],\ C_2(k) = [0\ 2],\ C_3(k) = [5\ 0],\ C_4(k) = [0\ 1].$$

Other parameters are set to be $\sigma_{ik} = 0.2$, $\bar{b}_k = 1$, and $\bar{v}_{ik} = 1$, for $k = 1, 2$ and $i = 1, 2, 3, 4$. Furthermore, the filter gains L_i are designed as follows

$$L_1(k) = [1\ 0],\ L_2(k) = [0\ 3],\ L_3(k) = [2\ 0],\ L_4(k) = [10\ 0].$$

such that the sensor network is distributively detectable. According Lemma 6.2.1, one can obtain

$$\Xi = \text{diag}\{0.3288, 0.6576, 0.1644, 0.6576\}$$

By choosing the strength $c = 2$, we can guarantee that the criteria (6.11) are fulfilled. For the sake of simulating the proposed event-based distributed filtering algorithm, we set the simulation step dt to be $0.01s$ and examine the triggering conditions at each step.

The simulation result is presented in Fig. 6.2, which depicts the trajectories of the mean-square error of the distributed filters under 1000 independent experiments. The result shows that the estimation error of the distributed filters interacting over Markovian switching topologies is exponentially mean-square bounded. Additionally, the number of events among 2000 steps is presented in Table 6.1, from which we confirm that the communication frequency has been greatly reduced.

6.4 Conclusions

This chapter has addressed the event-based distributed filtering problems over Markovian switching networks. To focus on specific components of the target state, we have introduced individual triggering conditions by utilizing individual thresholds for different components. We have shown that the exponential mean-square boundedness of distributed estimation error is always achieved if the sensor network is distributively detectable as a whole and the combined communication topology is strongly connected. Finally, the validity of the proposed filtering strategy has been illustrated via numerical simulation. We note that the proposed distributed filter is only valid when the nonlinear function $f(x(t), t)$ satisfies Assumption 6.1 and the measurement output $y(t)$ is linear. Then, a relevant research topic is to investigate the distributed filter design problem for general nonlinear systems based on the Takagi–Sugeno fuzzy model as [14]. Moreover, another future research topic would be the extension of our main results to more complex situations such as considering time delay or quantization effect in the communication process.

References

1. W. Ren, R. Beard, Consensus seeking in multi-agent systems under dynamically changing interaction topologies. IEEE Trans. Autom. Control **50**(5), 655–661 (2005)
2. F. Xiao, L. Wang, T. Chen, Connectivity preservation for multi-agent rendezvous with link failure. Automatica **48**(1), 25–35 (2012)
3. Y. Zhang, Y. Tian, Maximum allowable loss probability for consensus of multi-agent systems over random weighted lossy networks. IEEE Trans. Autom. Control **57**(8), 2127–2132 (2012)
4. K. You, Z. Li, L. Xie, Consensus condition for linear multi-agent systems over randomly switching topologies. Automatica **49**(10), 3125–3132 (2013)
5. X. Meng, T. Chen, Optimality and stability of event triggered consensus state estimation for wireless sensor networks, in *IEEE Proceedings 2014 American Control Conference*, (Portland, Oregon, USA 2014), pp. 3565–3570

6. Q. Liu, Z. Wang, X. He, D. Zhou, Event-based filtering with individual triggering thresholds in wireless sensor network: Distributed detectability analysis, in *IEEE Proceedings 34th Chinese Control Conference*, (Hangzhou, P.R. China, 2015), pp. 1734–1739
7. W. Yu, G. Chen, M. Cao, Consensus in directed networks of agents with nonlinear dynamics. IEEE Trans. Autom. Control **56**(6), 1436–1441 (2011)
8. W. Yu, G. Chen, M. Cao, J. Kurths, Second-order consensus for multi-agent systems with directed topologies and nonlinear dynamics. IEEE Trans. Syst. Man Cybern. Part B **40**(3), 881–891 (2010)
9. W. Li, Z. Wu, Output tracking of stochastic high-order nonlinear systems with Markovian switching. IEEE Trans. Autom. Control **58**(6), 1585–1590 (2013)
10. X. Mao, Exponential stability of stochastic delay interval systems with Markovian switching. IEEE Trans. Autom. Control **47**(10), 1604–1612 (2002)
11. S.M. Ross, *Stochastic Processes* (Wiley, New York, 1996)
12. T. Björk, Finite dimensional optimal filters for a class of Itô processes with jumping parameters. Stochastics-an Int. J. Probab. Stoch. Process. **4**(2), 167–183 (1980)
13. X. Liu, Z. Gao, Robust finite-time fault estimation for stochastic nonlinear systems with Brownian motions. J. Frankl. Inst. **354**(6), 2500–2523 (2017)
14. X. Liu, Z. Gao, Z.Q. Chen, Takagi-Sugeno fuzzy model based fault estimation and signal compensation with application to wind turbines. IEEE Trans. Ind. Electron. **64**(7), 5678–5689 (2017)

Chapter 7
Event-Based Recursive Distributed Filtering

An important practical problem with the wireless sensor networks is how to find distributed estimators or filters to extract the information about the state vectors of the target plants from observations contaminated with external disturbances. It is generally known that the traditional Kalman filter algorithm [1–5] is a recursive least mean square (LMS) one dealing with a single node and is optimal for linear systems with exact system models. On the other hand, to make use of the spatial information of the sensor nodes, distributed filtering problems have recently gained much research attention. Different from the traditional single node, in the distributed filtering schemes, the local filters estimate the system state based on the information not only from itself but also from its neighboring sensors according to the topology of networks. Up to now, many important results have been reported in the literature concerning the distributed filters and several effective strategies have been proposed, see, e.g., [6–11]. Among others, the H_∞ filter performance has gained particular research attention [6, 7, 9] due to its capability of ensuring certain worst-case performance with respect to external disturbances. In [12], the robust distributed state estimation problem has been addressed for stochastic sensor network systems based on the mean-square error analysis method. Furthermore, in [8], the distributed Kalman filter algorithm has been developed using dynamic consensus protocols with applications to the target tracking problems.

It is worth pointing out that most available results on distributed filtering problems have implicitly adopted the periodic communication strategy whose communication interval is designed *a priori* to guarantee desirable performance under the worst conditions. For some engineering systems where the communication bandwidth is not a concern, the periodic strategy could be an acceptable one for its simplicity in system analysis and design. Such a communication strategy, however, would probably lead to many unnecessary transmissions and therefore cause a waste of bandwidth resources. For example, in the case of wireless sensor networks communication constraints, frequent exchanges of signals would inevitably give rise to serious network

Q. Liu et al., *Stochastic Control and Filtering over Constrained Communication Networks*, Studies in Systems, Decision and Control 178,
https://doi.org/10.1007/978-3-030-00157-5_7

congestions, which might further induce network-related adverse phenomena such as packet losses and communication delays.

In recent years, there have been a growing number of results reported on event-based schedules whose aim is to decrease the unnecessary executions of the systems [13]. In an earlier work [14], the event-based sampling strategy has been compared with the time-driven one and it has been concluded that the former gave better performance for some simple systems. Later, much research efforts have been devoted to the applications of the event-based strategy to various engineering systems such as networked control systems [15–17] and multi-agent systems [18–20]. Unfortunately, when it comes to the state estimation problems, the available results in the literature have been scattered, most of which have focused on the practical implementation issues and there is a lack of adequate investigation on systematic analysis on the filtering performance. To be specific, a Send-on-Delta (SoD) regulation has been proposed in [21] for triggering mechanism whose effectiveness against time-driven sampling has been fully discussed. In [22], a modified Kalman filter has been investigated with intermittent measurement updates under the framework of event-based sampling. Another estimation problem with a SoD sampling strategy has been addressed in [23] without considering the communication delays and packet losses. In [24], the hybrid update strategy has been considered to reduce the estimation error based on the assumption that the difference between the present and the latest measurement values is limited within a bounded subset of the measurement space when no event is triggered. Besides, the event-based H_∞ filtering problems have been addressed in [25] for continuous-time systems with transmission delays.

Summarizing the above discussion, although the event-based estimation/filtering problem has stirred some initial research attention, the corresponding *distributed* filtering problem over *wireless sensor networks* has not yet been adequately investigated due probably to the difficulties in accommodating the topological information of the sensor networks. Note that the sparseness of the sensor network topology and the structure of the distributed filters add substantial challenges to the filter analysis and design, not to mention the difficulties brought from the event-based mechanism, especially when the filtering error variances are required to be locally minimized. As such, we are motivated to deal with the distributed filtering problem over wireless sensor networks according to the event-based strategy by employing a SoD concept. The employed SoD principle ensures that the executions are triggered if a certain signal exceeds a time-independent threshold defined as an important change of its value, and therefore the communication load and energy consumption in practice can be alleviated by avoiding unnecessary transmissions. With the proposed strategy, the sensor broadcasts its information to the neighboring nodes only when the function of its local innovation exceeds a fixed time-independent threshold. Obviously, our proposed strategy would reduce the number of transmissions over the wireless communication channel, thereby saving the resource.

The main contributions can be highlighted as follows: (1) a distributed filter is proposed for discrete time-varying systems in the framework of a novel event-based communication protocol; (2) a new event indicator variable is introduced to reflect the triggering information in the filter analysis with the hope to reduce possible conservatism in the filter analysis; (3) an upper bound of the filtering error variance is obtained by a recursive algorithm; and (4) the obtained upper bound is minimized at each iteration with proper filter parameters computed via a simplified matrix approach.

7.1 Problem Formulation and Preliminaries

Consider a discrete linear time-varying system described by the following state-space model:

$$x(k+1) = A_k x(k) + B_k w(k), \tag{7.1}$$

where $x(k) \in \mathbb{R}^{n_x}$ is the system state and $w(k) \in \mathbb{R}^{n_w}$ is the sequence of process noises. A_k, B_k are known system matrices of appropriate dimensions.

In this chapter, a sensor network consisting of n sensor nodes is investigated to measure the output of the target plant. We denote the topology of the network by a fixed directed graph $\mathcal{G} = (\mathcal{V}, \mathcal{E}, \mathcal{H})$ of order n with the set of nodes $\mathcal{V} = \{1, 2, \ldots, n\}$, the set of edges $\mathcal{E} \subseteq V \times V$, and the weighted adjacency matrix $\mathcal{H} = [a_{ij}]$. The weighted adjacency matrix of the graph is a matrix with nonnegative elements a_{ij} satisfying the property $a_{ij} > 0 \Longleftrightarrow (i, j) \in \mathcal{E}$, which means that the ith node can receive the information from the jth node. In this case, we can say that the ith nodes are the neighbor of the jth node. Moreover, all the neighbors of node i plus the node itself are denoted by the set as $N_i \triangleq \{j \in \mathcal{V} | (i, j) \in \mathcal{E}\}$.

For the ith sensor node, the model is described by

$$y_i(k) = C_{i,k} x(k) + D_{i,k} v(k), \tag{7.2}$$

where $y_i(k) \in \mathbb{R}^{n_y}$ stands for the measurement output received by sensor i, and $v(t) \in \mathbb{R}^{n_v}$ is the sequence of measurement noises. $C_{i,k}$, $D_{i,k}$ are known matrices of appropriate dimensions.

In this chapter, the following assumptions are made.

Assumption 7.1 $w(k)$ and $v(k)$ are mutually uncorrelated zero-mean Gaussian white noise sequences with respective covariances $Q_k > 0$ and $R_k > 0$.

Assumption 7.2 The initial state $x(0)$ has the mean $\bar{x}_0$ and covariance P_0, and is uncorrelated with both $w(k)$ and $v(k)$.

7.1.1 Traditional Distributed Filter Structure

Up to now, a large number of results have been developed with respect to the distributed filtering problems over the sensor networks. The traditional distributed filter structures have the general form as follows:

$$\hat{x}_i(k+1) = A_k\hat{x}_i(k) + \sum_{j\in N_i} a_{ij}G_{ij,k}r_j(k), \tag{7.3}$$

for $i = 1, 2, \ldots, n$, where $\hat{x}_i(k) \in \mathbb{R}^{n_x}$ is the estimation of the plant state in the ith sensor node. $G_{ij,k}$ is the filter parameter to be designed. Besides, $r_i(k) = y_i(k) - C_{i,k}\hat{x}_i(k)$ is the innovation sequence exchanged via the network.

So far, to the best of the authors' knowledge, almost all established results for distributed estimation problems basically assume that the sensor nodes should broadcast their local information at every periodic sampling instant. Such a scheme, however, will probably lead to a great deal of unnecessary data flow passing through the network, thereby inevitably aggravating the burden of communication network and increasing the power consumption. In order to significantly improve the network utilization efficiency, in the following, we propose to abandon the traditional periodic paradigm. Instead, we consider a novel event-based mechanism to suitably represent how sensors communicate with each other in a highly efficient way. The main advantage of the modified event-based mechanism lies in its capability of broadcasting the important messages only rather than all messages.

7.1.2 Event-Based Distributed Filter Structure

For the purpose of characterizing the SoD triggering mechanism, it is necessary to define event generator functions $f_i(.,.) : \mathbb{R}^{n_y} \times \mathbb{R} \to \mathbb{R}$ $(i = 1, \ldots, n)$ as follows:

$$f_i(e_i(k), \delta_i) = e_i^T(k)e_i(k) - \delta_i. \tag{7.4}$$

Here, $e_i(k) = r_i^t(k) - r_i(k)$ where $r_i^t(k)$ is the broadcast innovation at latest event time, $r_i(k)$ is the innovation sequence as defined before, and δ_i is a positive scalar. The executions are triggered as long as the condition $f_i(e_i(k), \delta_i) > 0$ is satisfied. Therefore, the sequence of the event-triggering instants $0 \le s_0^i \le s_1^i \le \cdots \le s_l^i \le \cdots$ is determined iteratively by

$$s_{l+1}^i = \min\{k \in \mathbb{N} | k > s_l^i, f_i(e_i(k), \delta_i) > 0\}. \tag{7.5}$$

Bearing in mind the event-based communication strategy, we can solve the distributed filtering problem via the following steps. First, the sensors synchronically measure the system state according to a time sequence $\{kh\}$ $(k = 0, 1, \ldots, \infty)$ where h is the sampling interval of the sensor nodes. Second, the event generator constructed in each sensor system only utilizes its local messages to check whether the triggering rule is fulfilled. Third, if the triggering condition is satisfied, the sensor node transmits the estimation information, namely, $r_i(k)$, to its adjacent nodes via wireless networks. Here, all the sensor nodes keep monitoring the wireless network and, as soon as receiving broadcasts or triggering predefined events, the corresponding sensor nodes update the input of local filters.

Remark 7.1 From the event generator function (7.4)–(7.5), it is indicated that once the variation of the difference $e_i(k)$ is intense enough and overweighs a certain threshold, the node has to inform all its neighboring sensor nodes of the newest messages so as to keep a satisfactory estimation performance of the wireless sensor network.

Remark 7.2 Note that the threshold δ_i is a parameter that regulates the triggering frequency. The smaller threshold means the higher frequency of event triggering. When $\delta_i = 0$ $(i = 1, 2, \ldots, n)$, the event-based approach reduces to a time-driven one, which requires the sensor nodes to broadcast their information at every sampling instant.

For simplicity, let us consider the time period $k \in [s_l^i, s_{l+1}^i)$. In the sequel, the event-based filter structures are adopted as follows:

$$\hat{x}_i(k+1) = A_k\hat{x}_i(k) + \sum_{j\in N_i} a_{ij}G_{ij,k}r_j^t(k), \tag{7.6}$$

for $i = 1, 2, ..., n,\ \ k \in [s_l^i, s_{l+1}^i)$, where $G_{ij,k}$ is the filter parameter to be designed.

It is worth mentioning that the transmission interval in the event-based schemes is usually larger than that in the periodic transmission schemes. Obviously, compared with existing filtering schemes, the event-based method has its inherently attractive features of reducing the burden of wireless network as well as the energy consumption of sensor systems, thereby helping to relieve the network congestion and increase the lifespan of the batteries in sensor nodes.

Remark 7.3 Due to the distributed nature of the filter algorithm, each sensor node could only access the local estimation and the messages of its neighbors at event-triggering instants. Hence, for sensor node i, the current filtering messages $r_j(k)$ $(j \in N_i)$ remain unavailable but, instead, we could take into account the latest updated values from its adjacent nodes, i.e., $r_j^t(k)$ $(j \in N_i)$.

The dynamics of the estimation error

$$\tilde{x}_i(k+1) = x_i(k+1) - \hat{x}_i(k+1),$$

can be obtained from (7.1), (7.2), and (7.6) as follows:

$$\begin{aligned}\tilde{x}_i(k+1) = {} & A_k\tilde{x}_i(k) - \sum_{j\in N_i} a_{ij}G_{ij,k}C_{j,k}\tilde{x}_j(k) + B_k w(k) \\ & - \sum_{j\in N_i} a_{ij}G_{ij,k}e_j(k) - \sum_{j\in N_i} a_{ij}G_{ij,k}D_{j,k}v(k).\end{aligned} \tag{7.7}$$

For the whole sensor network, we rewrite (7.7) in a more compact form and arrive at the following augmented filtering error system:

$$\begin{aligned}\tilde{x}(k+1) = (\mathcal{A}_k - \sum_{i=1}^{n} E_i G_k H_i \mathcal{C}_k)\tilde{x}(k) - \sum_{i=1}^{n} E_i G_k H_i e(k) \\ - \sum_{i=1}^{n} E_i G_k H_i \mathcal{D}_k \vartheta(k) + \mathcal{B}_k \omega(k),\end{aligned} \tag{7.8}$$

where

$$\begin{aligned}&\tilde{x}(k) = \mathrm{col}_n\{\tilde{x}_i(k)\}, \quad e(k) = \mathrm{col}_n\{e_i(k)\}, \quad \omega(k) = \mathrm{col}_n\{w(k)\}, \\ &\vartheta(k) = \mathrm{col}_n\{v(k)\}, \quad G_k = \{\, G_{ij,k} \,\}_{n\times n}, \quad \mathcal{A}_k = \mathrm{diag}_n\{A_k\}, \\ &\mathcal{B}_k = \mathrm{diag}_n\{B_k\}, \quad \mathcal{C}_k = \mathrm{diag}_n\{C_{i,k}\}, \quad \mathcal{D}_k = \mathrm{diag}_n\{D_{i,k}\}, \\ &H_i = \mathrm{diag}\{a_{i1}I, \ldots, a_{in}I\}, \quad E_i = \mathrm{diag}\{\underbrace{0, \ldots, 0}_{i-1}, I, \underbrace{0, \ldots, 0}_{n-i}\}.\end{aligned}$$

The aim of the addressed distributed filtering problem is to design the filter parameters, $G_{ij,k}$, in the filter (7.6) such that the filtering error covariance is bounded and such a bound is subsequently minimized iteratively by means of Riccati-like difference equations.

7.2 Main Results

In this section, we will first obtain an upper bound for the estimation error covariance of the system (7.8) and then look for an appropriate filter parameter G_k to minimize such an upper bound at each time-step k.

For presentation convenience, we denote

$$\begin{aligned}P_0 \triangleq \mathbb{E}\left\{\tilde{x}(0)\tilde{x}^T(0)\right\}, \quad P_k \triangleq \mathbb{E}\left\{\tilde{x}(k)\tilde{x}^T(k)\right\}, \\ \mathcal{Q}_k \triangleq \mathbb{E}\left\{\omega(k)\omega^T(k)\right\}, \quad \mathcal{R}_k \triangleq \mathbb{E}\left\{\vartheta(k)\vartheta^T(k)\right\}.\end{aligned}$$

Before proceeding further, we need to introduce the following lemma that will be utilized in the subsequent analysis.

Lemma 1 *For $0 \le k \le N$, suppose that $X = X^T \ge 0$, $Y = Y^T \ge 0$ and $\mathcal{X}_k(.) : \mathbb{R}^{nn_x \times nn_x} \to \mathbb{R}^{nn_x \times nn_x}$. If*

$$\mathcal{X}_k(X) \le \mathcal{X}_k(Y), \quad \forall X \le Y, \tag{7.9}$$

then the solutions W_{k+1} and M_{k+1} to the following difference equations

$$W_{k+1} = \mathcal{X}_k(W_k), \quad M_{k+1} \le \mathcal{X}_k(M_k), \quad M_0 = W_0 \tag{7.10}$$

satisfy

$$M_{k+1} \leq W_{k+1}. \tag{7.11}$$

Proof It follows from (7.10) that $M_0 \leq W_0$. Assuming, inductively, that $M_k \leq W_k$, we have

$$M_{k+1} \leq \mathcal{X}_k(M_k) \leq \mathcal{X}_k(W_k) = W_{k+1}. \tag{7.12}$$

The inductive hypothesis implies that $M_{k+1} \leq W_{k+1}$ is always true, which ends the proof.

Before deriving the upper bound for the filtering error variance, let us define the event indicator variables $\beta_i(k)$, which takes binary values 0 and 1, as follows: $\beta_i(k) = 0$ if the event generator conditions are satisfied at the current instant k for node i, while $\beta_i(k) = 1$ if no event is triggered. Furthermore, denote

$$\tilde{\beta}_k = \text{diag}_n\{\beta_i(k)I_{n_v}\},\ \bar{\beta}_k = I_{nn_v} - \text{diag}_n\{\beta_i(k)I_{n_v}\}, \tag{7.13}$$

which, according to the triggering condition (7.4)–(7.5), should be available at the current instant k for the purpose of calculating the filtering error covariance at the instant $k+1$.

Theorem 1 *Consider the linear time-varying system (7.1) with the distributed filters (7.6) and event generator condition (7.4)–(7.5). Let α be a positive scalar. For $0 \leq k \leq N-1$, assume that there exist two sets of real-valued matrices $\Xi_k > 0$ and G_k satisfying the following Riccati-like difference equation with the initial condition $\Xi_0 = P_0 \geq 0$:*

$$\begin{aligned}
\Xi_{k+1} &= \mathcal{X}_k(\Xi_k) \\
&\triangleq (1+\alpha)\Big(\mathcal{A}_k - \sum_{i=1}^{n} E_i G_k H_i \mathcal{C}_k\Big)\Xi_k\Big(\mathcal{A}_k - \sum_{i=1}^{n} E_i G_k H_i \mathcal{C}_k\Big)^T \\
&\quad + \Delta(1+\alpha^{-1})\Big(\sum_{i=1}^{n} E_i G_k H_i\Big)\Big(\sum_{i=1}^{n} E_i G_k H_i\Big)^T + \mathcal{B}_k \mathcal{Q}_k \mathcal{B}_k^T \\
&\quad + \Big(\sum_{i=1}^{n} E_i G_k H_i \mathcal{D}_k\Big)(\bar{\beta}_k \mathcal{R}_k \bar{\beta}_k - \tilde{\beta}_k \mathcal{R}_k \tilde{\beta}_k)\Big(\sum_{i=1}^{n} E_i G_k H_i \mathcal{D}_k\Big)^T,
\end{aligned} \tag{7.14}$$

where $\Delta = \sum_{i=1}^{n} \delta_i$. Then, the matrix Ξ_k is the upper bound of the estimation error covariance matrix P_k, that is, $\Xi_k \geq P_k$.

Proof The estimation error covariance at time-step $k+1$ is computed as follows:

$$
\begin{aligned}
P_{k+1} &= \mathbb{E}\big[\tilde{x}(k+1)\tilde{x}^T(k+1)\big] \\
&= \Big(\mathcal{A}_k - \sum_{i=1}^{n} E_i G_k H_i \mathcal{C}_k\Big)\mathbb{E}\big\{\tilde{x}(k)\tilde{x}^T(k)\big\}\Big(\mathcal{A}_k - \sum_{i=1}^{n} E_i G_k H_i \mathcal{C}_k\Big)^T \\
&\quad + \Big(\sum_{i=1}^{n} E_i G_k H_i\Big)\mathbb{E}\big\{e(k)e^T(k)\big\}\Big(\sum_{i=1}^{n} E_i G_k H_i\Big)^T \\
&\quad + \Big(\sum_{i=1}^{n} E_i G_k H_i \mathcal{D}_k\Big)\mathbb{E}\big\{\vartheta(k)\vartheta^T(k)\big\}\Big(\sum_{i=1}^{n} E_i G_k H_i \mathcal{D}_k\Big)^T \\
&\quad + \mathcal{B}_k\mathbb{E}\big\{\omega(k)\omega^T(k)\big\}\mathcal{B}_k^T + \mathcal{P}_k + \mathcal{P}_k^T + \mathcal{L}_k + \mathcal{L}_k^T \\
&\quad + \mathcal{R}_k + \mathcal{R}_k^T + \mathcal{Q}_k + \mathcal{Q}_k^T + \mathcal{T}_k + \mathcal{T}_k^T + \mathcal{S}_k + \mathcal{S}_k^T,
\end{aligned}
\tag{7.15}
$$

where

$$
\begin{aligned}
\mathcal{P}_k &= -\Big(\mathcal{A}_k - \sum_{i=1}^{n} E_i G_k H_i \mathcal{C}_k\Big)\mathbb{E}\big\{\tilde{x}(k)e^T(k)\big\}\Big(\sum_{i=1}^{n} E_i G_k H_i\Big)^T, \\
\mathcal{L}_k &= -\Big(\mathcal{A}_k - \sum_{i=1}^{n} E_i G_k H_i \mathcal{C}_k\Big)\mathbb{E}\big\{\tilde{x}(k)\vartheta^T(k)\big\}\Big(\sum_{i=1}^{n} E_i G_k H_i \mathcal{D}_k\Big)^T, \\
\mathcal{R}_k &= \Big(\mathcal{A}_k - \sum_{i=1}^{n} E_i G_k H_i \mathcal{C}_k\Big)\mathbb{E}\big\{\tilde{x}(k)\omega^T(k)\big\}\mathcal{B}_k^T, \\
\mathcal{Q}_k &= \Big(\sum_{i=1}^{n} E_i G_k H_i\Big)\mathbb{E}\big\{e(k)\vartheta^T(k)\big\}\Big(\sum_{i=1}^{n} E_i G_k H_i \mathcal{D}_k\Big)^T, \\
\mathcal{T}_k &= -\Big(\sum_{i=1}^{n} E_i G_k H_i\Big)\mathbb{E}\big\{e(k)\omega^T(k)\big\}\mathcal{B}_k^T, \\
\mathcal{S}_k &= -\Big(\sum_{i=1}^{n} E_i G_k H_i \mathcal{D}_k\Big)\mathbb{E}\big\{\vartheta(k)\omega^T(k)\big\}\mathcal{B}_k^T.
\end{aligned}
\tag{7.16}
$$

It follows immediately from

$$
\begin{aligned}
&\mathbb{E}\big\{\tilde{x}(k)\vartheta^T(k)\big\} = 0, \quad \mathbb{E}\big\{\tilde{x}(k)\omega^T(k)\big\} = 0, \\
&\mathbb{E}\big\{e(k)\omega^T(k)\big\} = 0, \quad \mathbb{E}\big\{\vartheta(k)\omega^T(k)\big\} = 0,
\end{aligned}
\tag{7.17}
$$

that $\mathcal{L}_k = 0$, $\mathcal{R}_k = 0$, $\mathcal{T}_k = 0$, $\mathcal{S}_k = 0$. However, different from the traditional filtering problems, the expectations of some cross terms in $\mathcal{P}_k$ and $\mathcal{Q}_k$ are no longer zero, which requires further investigations.

First, recall the definition of the gap $e_i(k) = r_i^t(k) - r_i(k)$. For the case that the current sampling instant happens to be the event-triggering instant of the ith node, namely, $k = s_l^i$, it is obvious that $e_i(k) = 0$ and therefore $\mathbb{E}\{e_i(k)v^T(k)\} = 0$. Otherwise, as $e_i(k) = r_i^t(k) - (C_{i,k}\tilde{x}(k) + D_{i,k}v(k))$, we have

$$\mathbb{E}\big\{e_i(k)v^T(k)\big\} = \mathbb{E}\Big\{[r_i^t(k) - (C_{i,k}\tilde{x}(k) + D_{i,k}v(k))]v^T(k)\Big\} = -D_{i,k}R_k.$$

It can now be concluded that

$$\mathbb{E}\big\{e(k)\vartheta^T(k)\big\} = -\mathcal{D}_k\tilde{\beta}_k\mathcal{R}_k, \tag{7.18}$$

where $\mathcal{R}_k = [R_k]_{n\times n}$. Subsequently, we have

$$\mathcal{Q}_k + \mathcal{Q}_k^T = \Big(\sum_{i=1}^{n} E_iG_kH_i\mathcal{D}_k\Big)\Big[\bar{\beta}_k\mathcal{R}_k\bar{\beta}_k - \tilde{\beta}_k\mathcal{R}_k\tilde{\beta}_k - \mathcal{R}_k\Big]\Big(\sum_{i=1}^{n} E_iG_kH_i\mathcal{D}_k\Big)^T. \tag{7.19}$$

In the next step, by using the elementary inequality $\big(\alpha^{\frac{1}{2}}M - \alpha^{-\frac{1}{2}}N\big)\big(\alpha^{\frac{1}{2}}M - \alpha^{-\frac{1}{2}}N\big)^T \geqslant 0$ where M and N are matrices with compatible dimensions, it follows from (7.16) that

$$\begin{aligned}\mathcal{P}_k + \mathcal{P}_k^T \le\ & \alpha\Big(\mathcal{A}_k - \sum_{i=1}^{n} E_iG_kH_i\mathcal{C}_k\Big)\mathbb{E}\big\{\tilde{x}(k)\tilde{x}^T(k)\big\}\Big(\mathcal{A}_k - \sum_{i=1}^{n} E_iG_kH_i\mathcal{C}_k\Big) \\ & + \alpha^{-1}\Big(\sum_{i=1}^{n} E_iG_kH_i\Big)\mathbb{E}\big\{e(k)e^T(k)\big\}\Big(\sum_{i=1}^{n} E_iG_kH_i\Big)^T.\end{aligned} \tag{7.20}$$

In addition, under the event-based strategy, the gap $e_i(k)$ would be immediately reset to zero if the triggering condition is fulfilled. Consequently, the following inequality is always satisfied:

$$e^T(k)e(k) \le \Delta, \tag{7.21}$$

where $\Delta = \sum_{i=1}^{n}\delta_i$. Applying the properties of matrix operations, we obtain

$$e(k)e^T(k) \le \|e(k)\|^2 I = e^T(k)e(k)I \le \Delta I, \tag{7.22}$$

and hence

$$\mathbb{E}\{e(k)e^T(k)\} \le \Delta I. \tag{7.23}$$

Now, it follows from (7.17), (7.19), (7.20), and (7.23) that

$$P_{k+1} \le \mathcal{X}_k(P_k). \tag{7.24}$$

Define a positive semidefinite matrix $\Xi_0 = P_0$, and Ξ_{k+1} can be calculated iteratively by the Riccati-like difference equation:

$$\Xi_{k+1} = \mathcal{X}_k(\Xi_k). \tag{7.25}$$

Obviously, the conditions in Lemma 1 are satisfied. Therefore, for all $0 \leq k \leq N$, we have

$$P_{k+1} \leq \Xi_{k+1}, \tag{7.26}$$

and the proof is now complete.

Remark 7.4 It is noticed that, in (7.14), the term $(\sum_{i=1}^{n} E_i G_k H_i)(\sum_{i=1}^{n} E_i G_k H_i)^T \geq 0$. Therefore, the upper bound will increase as the threshold Δ increases. From the engineering viewpoint, a larger threshold would lead to a fewer number of transmissions over the network, which implies that the threshold Δ does have a major impact on the tradeoffs between the filtering performance and the sensor data transmission rate.

Remark 7.5 In the recursive calculation of the *performance index*, namely, the filtering error covariance, the triggering information at the time instant k is explicitly exploited in (7.14) in terms of the binary variable $\beta_i(k)$. This would definitely help reduce the conservatism and tighten the upper bound. Comparing to the traditional event-triggering mechanism where only the stability is the concern, the introduction of such a binary variable $\beta_i(k)$ would play an important role in estimating the filtering performance (i.e., the error covariance).

Remark 7.6 In the case of threshold $\delta_i = 0$, it is not difficult to see that the triggering rules are always fulfilled for every sensor nodes and the indicator $\tilde{\beta}_k$ is a zero matrix with appropriate dimensions. Consequently, the Riccati-like difference equation (7.14) would reduce to the recursion of the estimation error covariance for the traditional distributed filtering problem.

We are now ready to minimize the upper bound Ξ_k on each recursion by appropriately designing the filter parameters. For this purpose, the following lemma is useful.

Lemma 2 *For matrices M, X, N, L with compatible dimensions, the following are true:*

$$\begin{aligned}
&\frac{\partial}{\partial X}\mathrm{tr}(MX^T) = M, \quad \frac{\partial}{\partial X}\mathrm{tr}(XM) = M^T, \\
&\frac{\partial}{\partial X}\mathrm{tr}(MXN) = M^T N^T, \quad \frac{\partial}{\partial X}\mathrm{tr}(MX^T N) = NM, \\
&\frac{\partial}{\partial X}\mathrm{tr}(MXNX^T L) = M^T L^T X N^T + LMXN.
\end{aligned} \tag{7.27}$$

Furthermore, for any symmetric matrix P, the following holds:

$$\frac{\partial}{\partial X}\text{tr}\{(MXN)P(MXN)^T)\} = 2M^TMXNPN^T. \tag{7.28}$$

For presentation simplicity, we denote

$$\begin{aligned}
\mathcal{A}_k^{(i)} &= \{\underbrace{0,\ldots,0}_{i-1}, A_k, \underbrace{0,\ldots,0}_{n-i}\},\\
G_k &= \{G_{ij,k}\}_{n\times n},\ \ G_k^{(i)} = \left\{G_{i1,k}, G_{i2,k},\cdots, G_{in,k}\right\},\\
\mathcal{M}_{i,k} &= (1+\alpha)H_i\mathcal{C}_k\Xi_k(H_i\mathcal{C}_k)^T + \Delta(1+\alpha^{-1})H_iH_i,\\
&\quad + H_i\mathcal{D}_k(\bar{\beta}_k\mathcal{R}_k\bar{\beta}_k - \tilde{\beta}_k\mathcal{R}_k\tilde{\beta}_k)(H_i\mathcal{D}_k)^T,\\
\mathcal{N}_{i,k} &= (1+\alpha)\mathcal{A}_k^{(i)}\Xi_k(H_i\mathcal{C}_k)^T.
\end{aligned} \tag{7.29}$$

Furthermore, let $\bar{G}_k^{(i)}$ and $\bar{\mathcal{N}}_{i,k}$ be the simplified matrices by removing the jth ($j \notin N_i$) column from $G_k^{(i)}$ and $\mathcal{N}_{i,k}$, respectively. Also, let $\bar{\mathcal{M}}_{i,k}$ be a simplified matrix by removing both the jth ($j \notin N_i$) row and jth ($j \notin N_i$) column from $\mathcal{M}_{i,k}$.

Theorem 2 *Consider the linear time-varying system (7.1) with the distributed filters (7.6) and event generator condition (7.4)–(7.5). The solution Ξ_k to (7.14), which is an upper bound of the estimation error covariance matrix P_k, can be minimized at each iteration with the filter parameter $G_k = \{G_{ij,k}\}_{n\times n}$ given by*

$$\bar{G}_k^{(i)} = \bar{\mathcal{N}}_{i,k}\bar{\mathcal{M}}_{i,k}^{-1},\ \text{for all } 0 \le i \le n, 0 \le k \le N-1, \tag{7.30}$$

and $G_{ij,k} = 0$ for $j \notin N_i$. Here, $\bar{G}_k^{(i)}$, $\bar{\mathcal{N}}_{i,k}$, and $\bar{\mathcal{M}}_{i,k}$ are defined right after (7.29).

Proof According to (7.14), one has

$$\begin{aligned}
&\text{tr}(\Xi_{k+1})\\
&= (1+\alpha)\text{tr}\Big\{\mathcal{A}_k\Xi_k\mathcal{A}_k^T + \Big(\sum_{i=1}^n E_iG_kH_i\mathcal{C}_k\Big)\Xi_k\Big(\sum_{i=1}^n E_iG_kH_i\mathcal{C}_k\Big)^T\\
&\quad - \mathcal{A}_k\Xi_k\Big(\sum_{i=1}^n E_iG_kH_i\mathcal{C}_k\Big)^T - \Big(\sum_{i=1}^n E_iG_kH_i\mathcal{C}_k\Big)\Xi_k\mathcal{A}_k^T + \mathcal{B}_k\mathcal{Q}_k\mathcal{B}_k^T\\
&\quad + \Big(\sum_{i=1}^n E_iG_kH_i\mathcal{D}_k\Big)\Big(\bar{\beta}_k\mathcal{R}_k\bar{\beta}_k - \tilde{\beta}_k\mathcal{R}_k\tilde{\beta}_k\Big)\Big(\sum_{i=1}^n E_iG_kH_i\mathcal{D}_k\Big)^T\\
&\quad + \Delta(1+\alpha^{-1})\Big(\sum_{i=1}^n E_iG_kH_i\Big)\Big(\sum_{i=1}^n E_iG_kH_i\Big)^T\Big\}.
\end{aligned}$$

Resorting to the properties of trace, we have

$$\mathrm{tr}\left\{E_i M\, \Xi_k (E_j N)^T\right\} = 0, \ \text{if } i \neq j,$$

where M and N are matrices with appropriate dimensions. Next, taking the partial derivation of the trace of the matrix Ξ_{k+1} with respect to the parameters G_k, and utilizing Eqs. (7.27) and (7.28), we arrive at

$$\begin{aligned}
&\frac{\partial}{\partial G_k} tr\{\Xi_{k+1}\} \\
&= 2(1+\alpha)\Big\{\sum_{i=1}^{n} E_i G_k H_i \mathcal{C}_k \Xi_k (H_i \mathcal{C}_k)^T - \sum_{i=1}^{n} E_i \mathcal{A}_k \Xi_k (H_i \mathcal{C}_k)^T\Big\} \\
&+ 2\Delta(1+\alpha^{-1})\sum_{i=1}^{n} E_i G_k H_i H_i + 2\sum_{i=1}^{n} E_i G_k H_i \mathcal{D}_k (\bar{\beta}_k \mathcal{R}_k \bar{\beta}_k - \tilde{\beta}_k \mathcal{R}_k \tilde{\beta}_k)(H_i \mathcal{D}_k)^T.
\end{aligned}$$

By solving the function $\frac{\partial}{\partial G_k} tr\{\Xi_{k+1}\} = 0$, we can get the optimal gain G_k that minimizes $tr\{\Xi_{k+1}\}$. Subsequently, as $G_k^{(i)}$ and $\mathcal{A}_k^{(i)}$ represent, respectively, the ith row of matrices G_k and the ith row of $\mathcal{A}_k$, we have

$$G_k^{(i)} \mathcal{M}_{i,k} = \mathcal{N}_{i,k}. \tag{7.31}$$

Recalling the definition $H_i = \mathrm{diag}\{a_{i1}I, a_{i2}I, , \ldots, a_{in}I\}$ and the relation $a_{ij} = 0$ ($j \notin N_i$), one can remove the corresponding zero rows and zero columns from $\mathcal{M}_{i,k}$ and the corresponding zero columns from $\mathcal{N}_{i,k}$ to obtain the simplified matrices $\bar{\mathcal{M}}_{i,k}$ and $\bar{\mathcal{N}}_{i,k}$. Let $\mathcal{M}_{i,k} = \{\tilde{M}_{ab,k}\}_{n\times n}$ and $\mathcal{N}_{i,k} = \{\tilde{N}_{b,k}\}_{1\times n}$. If $b \in N_i$, it follows that

$$\sum_{j=1}^{n} G_{ij,k} \tilde{M}_{jb,k} = \tilde{N}_{b,k}. \tag{7.32}$$

In other cases, the above equation is always satisfied since both sides of it are zero. Because $G_{ij,k}, j \notin N_i$ are chosen as zero matrices of appropriate dimensions and the corresponding rows of $\mathcal{M}_{i,k}$ are also zeros, it can be seen that

$$\bar{G}_k^{(i)} \bar{\mathcal{M}}_{i,k}^{(b')} = \bar{\mathcal{N}}_{i,k}^{(b')}, \tag{7.33}$$

where $\bar{G}_k^{(i)}$ is a simplified matrix of $G_k^{(i)}$ by removing $G_{ij,k}, j \notin N_i$, b' is the new column index of the simplified matrix, and $\bar{\mathcal{M}}_{i,k}^{(b')}$, $\bar{\mathcal{N}}_{i,k}^{(b')}$ are the b'th column of matrices $\bar{\mathcal{M}}_{i,k}, \bar{\mathcal{N}}_{i,k}$, respectively. It is not difficult to verify that the matrix $\bar{\mathcal{M}}_{i,k}$ is positive definite and the filter parameters can be calculated as follows:

$$\bar{G}_k^{(i)} = \bar{\mathcal{N}}_{i,k} \bar{\mathcal{M}}_{i,k}^{-1}, \tag{7.34}$$

which completes the proof.

Remark 7.7 In Theorem 2, a novel matrix simplification technique is developed to handle the challenges resulting from the sparseness of the sensor network topology and the preserving structure of the distributed filters. By using the employed SoD principle, the executions are triggered when a certain signal exceeds a time-independent threshold defined as an important change of its value, and therefore the communication load and energy consumption in practice can be alleviated by avoiding unnecessary transmissions.

Remark 7.8 In the time-invariant case, all system parameters are constant matrices and the boundedness problem of Ξ_k can be dealt with by examining the Schur stability of matrix $(1+\alpha)^{1/2}(\mathcal{A}-\sum_{i=1}^{n} E_i G H_i \mathcal{C})$. In this case, by properly designing the filter parameter, the estimate error can be guaranteed to be bounded.

Remark 7.9 In this chapter, the distributed filtering problem is solved for a class of discrete time-varying systems with an event-based communication mechanism. By using an event indicator variable, the triggering information is included in the Riccati-like difference equation whose solution serves as an upper bound for the filtering error covariance which is, subsequently, minimized by appropriately designing the filter parameters iteratively. Note that the main result established in Theorem 2 contains all the information about the system parameters, the topology information as well as the triggering information.

7.3 An Illustrative Example

To illustrate the validity of the proposed filter design strategy, we consider a second-order system (7.1) where $x(k) = (x_1(k)\ x_2(k))^T \in \mathbb{R}^2$, the random noise $w(k)$ is zero-mean Gaussian sequences with covariance $Q_k = 1$, and

$$A_k = \begin{bmatrix} 0.98 + 0.05\sin(0.12k) & 0.4 \\ 0.15 & -0.75 \end{bmatrix},$$
$$B_k = \begin{bmatrix} 0.16\ 0.18 \end{bmatrix}^T.$$

Suppose the initial value of the state $x(0)$ is uniformly distributed over the region $[-6, 6]$ so that $\mathbb{E}\{x(0)\} = [0\ 0]$.

The sensor network is represented by a directed graph $\mathcal{G} = (\mathcal{V}, \mathcal{E}, \mathcal{A})$ where the set of nodes $\mathcal{V} = \{1, 2, 3, 4\}$, the set of edges $\mathcal{E} = \{(1, 1), (1, 2), (2, 1), (2, 2), (2, 3), (3, 1), (3, 3), (4, 1), (4, 4)\}$, and the adjacency elements associated with the edges of the graph are $a_{ij} = 1$. The dynamics of the sensor nodes is modeled as (7.2) with

$$\begin{aligned}
C_{1,k} &= [0.82,\ 0.62 + 0.05\cos(0.12k)],\\
C_{2,k} &= [0.75 + 0.05\sin(0.1k),\ 0.80],\\
C_{3,k} &= [0.74 + 0.05\sin(0.1k),\ 0.75 + 0.05\cos(0.1k)],\\
C_{4,k} &= [0.75,\ 0.65].
\end{aligned}$$

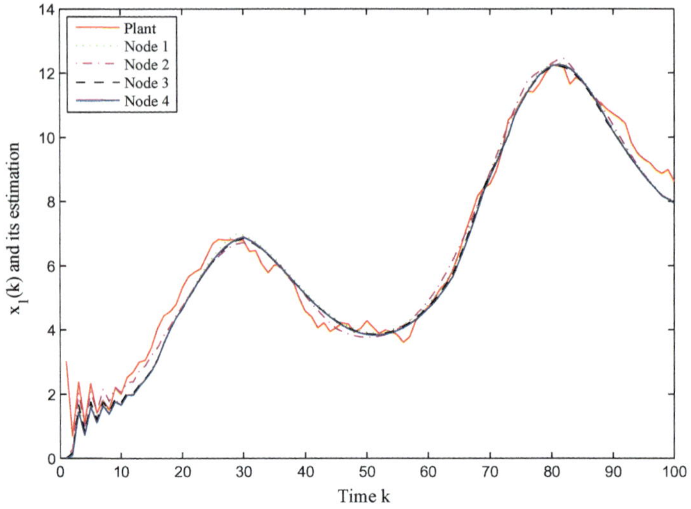

Fig. 7.1 The state x_1 and its estimation

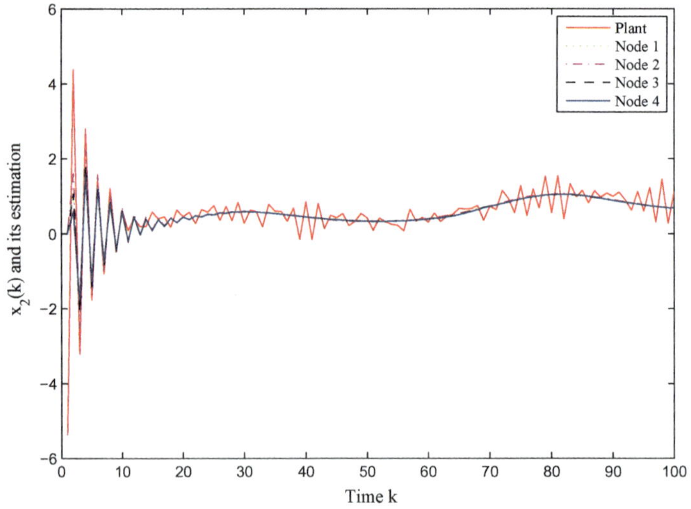

Fig. 7.2 The state x_2 and its estimation

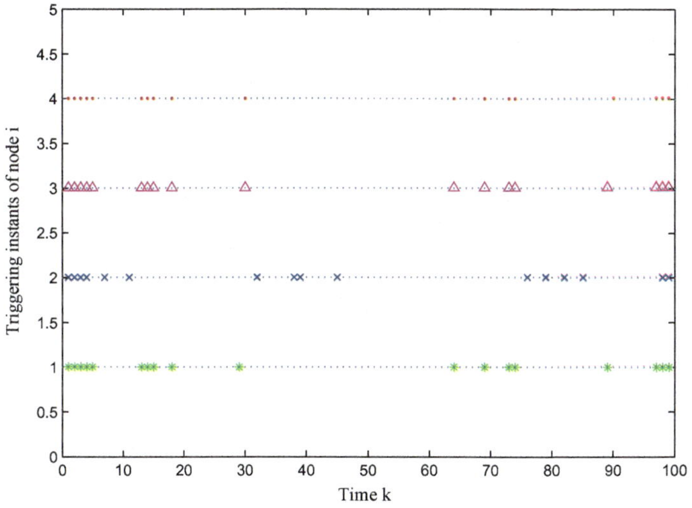

Fig. 7.3 The triggering sequence

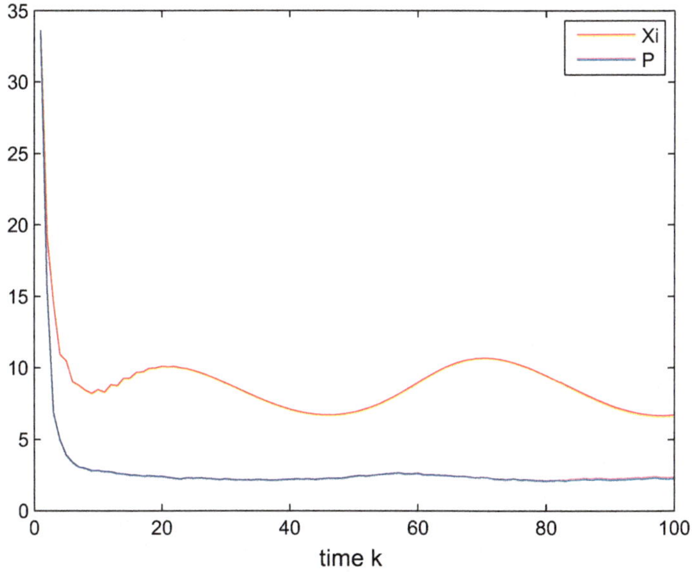

Fig. 7.4 The trace of error covariance and its upper boundary

In the simulation, let the measurement noise $v(k)$ be zero-mean Gaussian white noise sequences with covariance $R_k = 1$. Choose the thresholds $\delta_i = 0.4$ for $i = 1, 2, 3, 4$ and the scalar $\alpha = 0.1$. The initial estimations about the plant state are $\hat{x}_1(0) = \hat{x}_2(0) = \hat{x}_3(0) = \hat{x}_4(0) = [0\ 0]$. The trace of mean-square error for the esti-

Table 7.1 The influence of thresholds

$\sigma_i, i = 1, \ldots, 4$	0.1	0.2	0.4	0.8
Average communication rate	1.8908	1.3328	0.8160	0.5316

Table 7.2 The influence of noise covariances

(Q_k, R_k)	(0.5, 0.5)	(1,1)	(2,2)	(4,4)
$\sigma_i, i = 1, \ldots, 4$	0.1	0.2	0.4	0.8
Average communication rate	0.3920	0.8160	1.9164	2.8608

mation of the state is averaged in 500 runs of simulation. The suboptimal upper bound and corresponding filter parameters G_k can be calculated at each iteration according to (7.14) and (7.30).

Simulation results are shown in Figs. 7.1, 7.2, 7.3, and 7.4. In Figs. 7.1 and 7.2, the trajectories for the states and respective estimates are depicted, from which we can see that the proposed strategy has a favorable performance in estimating the state of a divergent system. Besides, compared with time-driven mechanism, the broadcast times are significantly reduced in Fig. 7.3, which embodies the superiority of the event-based one. To this end, Fig. 7.4 shows the upper bound Ξ_k of the error covariance matrix P_k.

To evaluate the influence of event thresholds and noise covariances on the communication rate (total number of communications in the network divided by running time), we have the following experimental results via 500 trials. The results are shown in Tables 7.1 and 7.2. Obviously, both the decrease in thresholds and the increase in noise covariances would contribute to a higher communication frequency.

7.4 Conclusions

In this chapter, a novel event-based distributed filter has been proposed in wireless sensor networks for the sake of reducing the sensor data transmission rate and the energy consumption. Basing on a SoD concept, each smart sensor node makes decisions independently about when the local messages have to be broadcast. By using a combination of trace properties, inductive method, and event-induced behaviors, we have computed the upper bound for the filtering error covariance recursively by solving a Riccati-like matrix equation. Subsequently, the distributed filter gain has been properly designed at every iteration to minimize such an upper bound. Finally, via a numerical example, we have demonstrated the effectiveness of the event-based communication protocol in reducing the number of transmissions over the network.

References

1. G. Welch, G. Bishop, An introduction to the Kalman filter, *Notes of ACM SIGGRAPH tutorial on the Kalman Filter, 2001*
2. M. Basin, P. Shi, D. Calderon-Alvarez, J. Wang, Central suboptimal $\mathcal{H}_\infty$ filter design for linear time-varying systems with state or measurement delay. Circuits Syst. Signal Process. **28**(2), 305–330 (2009)
3. R. Caballero-Aguila, A. Hermoso-Carazo, J.D. Jimenez-Lopez, J. Linares-Perez, S. Nakamori, Signal estimation with multiple delayed sensors using covariance information. Digit. Signal Process. **20**(2), 528–540 (2010)
4. H. Gao, T. Chen, J. Lam, A new delay system approach to network-based control. Automatica **44**(1), 39–52 (2008)
5. S. Sun, L. Xie, W. Xiao, Optimal full-order and reduced-order estimators for discrete-time systems with multiple packet dropouts. IEEE Trans. Signal Process. **56**(8), 4031–4038 (2008)
6. H. Dong, Z. Wang, H. Gao, Distributed H_∞ filtering for a class of Markovian jump nonlinear time-delay systems over lossy sensor networks. IEEE Trans. Indus. Electron. **60**(10), 4665–4672 (2013)
7. B. Shen, Z. Wang, Y.S. Hung, Distributed H_∞-consensus filtering in sensor networks with multiple missing measurements: the finite-horizon case. Automatica **46**(10), 1682–1688 (2010)
8. R. Olfati-Saber, Distributed Kalman filtering for sensor networks, in *Proceeding 46th IEEE Conference Decision and Control*, New Orleans, LA, USA (pp. 5492–5498, 2007)
9. D. Ding, Z. Wang, B. Shen, H. Shu, H_∞ state estimation for discrete-time complex networks with randomly occurring sensor saturations and randomly varying sensor delays. IEEE Trans. Neural Netw. **23**(5), 725–736 (2012)
10. J. Liang, B. Shen, H. Dong, J. Lam, Robust distributed state estimation for sensor networks with multiple stochastic communication delays. Int. J. Syst. Sci. **42**(9), 1459–1471 (2011)
11. S. Havedanloo, H.R. Karimi, Improving the performance metric of wireless sensor networks with clustering Markov chain model and multilevel fusion. Math. Problems Eng. Art. No. 783543 (2013)
12. M.V. Basin, J.J. Maldonado, H.R. Karimi, Mean-square filtering for polynomial system states confused with poisson noises over polynomial observations. Model. Identif. Control **32**(2), 47–55 (2011)
13. Q. Liu, Z. Wang, X. He, D.H. Zhou, A survey of event-based strategies on control and estimation. Syst. Sci. Control Eng. **2**(1), 90–97 (2014)
14. R. Zheng, J. Hou, N. Li, Power management and power control in wireless networks. Ad. Hoc Sensor. (Netw. Nova Science Publishers, New York, 2004), pp. 1–25
15. X. Wang, M. Lemmon, Decentralized event-triggered broadcasts over networked control systems. Hybrid Syst. Comput. Contro (Springer, Berlin, 2008) pp. 674–677
16. E. Garcia, P. Antsaklis, Model-based event-triggered control for systems with quantization and time-varying network delays. IEEE Trans. Autom. Control **58**(2), 422–434 (2013)
17. D. Yue, E. Tian, Q. Han, A delay system method for designing event-triggered controllers of networked control systems. IEEE Trans. Autom. Control **58**(2), 475–481 (2013)
18. J. Hu, G. Chen, H.X. Li, Distributed event-triggered tracking control of leader-follower multi-agent systems with communication delays. Kybernetika **47**(4), 630–643 (2011)
19. D.V. Dimarogonas, E. Frazzoli, K.H. Johansson, Distributed event-triggered control for multi-agent systems. IEEE Trans. Autom. Control **57**(5), 1291–1297 (2012)
20. X. Yin, D. Yue, Event-triggered tracking control for heterogeneous multi-agent systems with Markov communication delays. J. Frankl. Inst. **350**(5), 1312–1334 (2013)
21. M. Miskowicz, Send-on-delta concept: an event-based data reporting strategy. Sensors **6**(1), 49–63 (2006)
22. A. Le, R. McCann, Event-based measurement updating Kalman filter in network control systems, in *Proceedings IEEE Region 5 Technical Conference*, (Fayetteville, AR, USA, 2007) pp. 138–141

23. V.H. Nguyen, Y.S. Suh, Improving estimation performance in networked control systems applying the send-on-delta transmission method. Sensors **7**(10), 2128–2138 (2007)
24. Y.S. Suh, V.H. Nguyen, Y.S. Ro, Modified Kalman filter for networked monitoring systems employing a send-on-delta method. Automatica **43**(2), 332–338 (2007)
25. S. Hu, D. Yue, Event-based H_∞ filtering for networked system with communication delay. Signal Process. **92**(9), 2029–2039 (2012)

Chapter 8
A Resilient Approach to Distributed Recursive Filter Design

The state estimation or filtering problem has proven to be one of the fundamental issues in signal processing and control engineering, and a number of algorithms have been proposed in the literature, see, e.g., [1–10]. Accordingly, a core problem with the widespread applications of wireless sensor networks (WSNs) is to estimate the plant states based on noisy measurement outputs from distributed nodes. A seemingly natural way is to employ the traditional Kalman filters by establishing a fusion center in WSNs in order to collect *all* the measurements from the individual sensors and then process the measurements in a *global* sense. Unfortunately, due to the limited communication capability and energy supply, it might be impossible for the sensors to persistently forward the local messages to the fusion center. As such, the so-called *distributed* estimation scheme would be more preferable whose main idea is to estimate the plant states based on both the local and the neighboring information according to the topologies of WSNs. Recently, various types of consensus protocols have been proposed with an aim to improve the efficiency of the distributed computation and a rich body of literature has been available on the consensus-based distributed filtering strategies, see, e.g., the seminal work in [11].

As is well known, nonlinearities exist in almost all practical systems and the corresponding research on nonlinear control problems has served as one of the mainstream areas in systems and control communities. In certain noisy environments such as networked control systems, the nonlinearity disturbances may result from randomly fluctuated network conditions and/or communication constraints. In this case, the so-called stochastic uncertainties would become inevitable that might lead to serious degradation of system performance if not properly dealt with. So far, there has been a growing research interest in analysis and synthesis issues for the systems with stochastic nonlinearities. Some representative results have been reported in [12–14] and the references therein, where sophisticated models have been proposed to characterize the random occurrence of the nonlinearities through the statistics (typically the first- and second-order moments). On the other hand, in engineering practice, the

Q. Liu et al., *Stochastic Control and Filtering over Constrained Communication Networks*, Studies in Systems, Decision and Control 178,
https://doi.org/10.1007/978-3-030-00157-5_8

phenomenon of sensor degradation may occur randomly as well, which is caused by various factors ranging from sensor aging and sensor intermittent failure to transmission congestions, see [15]. Some research effort has been initiated on the estimation problem with sensor degradations, see, e.g., [16–18]. However, when it comes to the distributed estimation problems, the corresponding results have been very few, not to mention the case where the stochastic nonlinearities are also a concern. Such a situation gives the primary motivation for the present investigation.

Another critical issue of practical significance in state estimation problems is the resilience of the filters, which can be understood as the insensitivity against possible parameter variations/drifts in implementing the designed filters. In most previous works, the filter algorithms have been developed based on an implicit assumption that the designed filter gains can be *accurately* realized. Such an assumption is, unfortunately, not always true in practice due primarily to the finite resolution instrumentation during filter implementation and the round-off errors resulting from the fixed word length in numerical computation. Since the performance of filters can be extremely fragile to even tiny gain variations, it is of vital importance to design resilient filters capable of tolerating possible gain variations. In the past decade, much work has been done to ensure the resilience (or non-fragility) of the filters/estimators/controllers, see, e.g., [16, 17, 19–23]. For example, the problem of robust non-fragile Kalman filter design has been studied in [19] for a class of linear systems with norm-bounded uncertainties, and some new criteria have been provided to guarantee the mean-square stability in terms of the solutions to algebraic Riccati equations. The minimum variance state estimation problems have been considered in [16, 17] for linear and nonlinear systems with both sensor failures and gain perturbations in the case of centralized filtering.

Summarizing the above discussions, it can be concluded that there is a lack of systematic investigation on the distributed estimation problem for systems subject to stochastic nonlinearities, sensor degradation as well as filter gain perturbations over wireless sensor networks with a given topology. As such, the main purpose of this chapter is to shorten such a gap by designing distributed filters that are resilient to filter implementation errors and robust to sensor degradations. The main contributions can be highlighted as follows: (1) the system under consideration is quite general that covers stochastic nonlinearities and sensor degradation; (2) a resilient distributed filtering algorithm is proposed so as to mitigate the adverse effects induced by filter gain variations; (3) a matrix simplification approach is exploited in the filter design algorithm to overcome the difficulties resulting from the sparsity of the sensor networks; and (4) a criterion is established for the mean-square boundedness of the estimator errors for the designed time-varying distributed resilient filter.

The remainder of this chapter is organized as follows: Section 8.1 formulates the problem to be studied. Section 8.2 presents some preliminary results. Section 8.3 is concerned with the algorithm for designing the distributed resilient filters. Section 8.4 gives a sufficient condition for the boundedness of the estimation errors. A numerical example is provided in Sect. 8.5 and concluding remarks are drawn in Sect. 8.6.

8.1 Problem Formulation

8.1.1 Target Plant and Sensor Network

In this chapter, a sensor network consisting of n sensor nodes is exploited to measure the output of the target plant. We denote the topology of the network by a directed graph $G = (V, E, \mathcal{H})$ of order n with the set of nodes $V = \{1, 2, \ldots, n\}$, the set of edges $E \subseteq V \times V$, and the weighted adjacency matrix $\mathcal{H} = [a_{ij}]_{n\times n}$. The weighted adjacency matrix of the graph is a matrix with nonnegative elements a_{ij} satisfying the property $a_{ij} > 0 \Longleftrightarrow (i, j) \in E$, which means that the ith node can receive the information from the jth node. All the neighbors of node i plus the node itself are denoted by the set as $N_i \triangleq \{j \in V|(i, j) \in E\}$.

Consider the following discrete time-varying target plant with stochastic nonlinearities:

$$x(k+1) = A(k)x(k) + f(k, x(k), \xi(k)) + w(k), \tag{8.1}$$

where $x(k) \in \mathbb{R}^{n_x}$ is the state vector that cannot be measured directly, $f(k, x(k), \xi(k)) \in \mathbb{R}^{n_x}$ is the stochastic nonlinearities to be defined later, and $w(k) \in \mathbb{R}^{n_x}$ is a sequence of Gaussian random variables with zero-mean value and covariance matrix $Q(k) > 0$. $A(k)$ is a known time-varying matrix of appropriate dimensions. The initial condition $x(0)$ is assumed to obey a Gaussian distribution with mean μ_0 and covariance matrix Σ_0.

For the ith ($i = 1, 2, \ldots, n$) sensor node, the measurement is described by

$$y_i(k) = \gamma_i(k)C_i(k)x(k) + g_i(k, x(k), \zeta_i(k)) + v_i(k), \tag{8.2}$$

where $y_i(k) \in \mathbb{R}^{n_y}$ stands for the measurement information from sensor i and the measured noise $v_i(k) \in \mathbb{R}^{n_y}$ obeys a Gaussian distribution with zero-mean value and covariance matrix $R_i(k) > 0$. The variable $\gamma_i(k)$ accounting for the sensor gain degradation has the probability density function $p_k^i(.)$ on the interval $[0, 1]$ with mean $\bar{\gamma}_i(k)$ and variance $\sigma^2_{\gamma_i}(k)$. $C_i(k)$ is a known time-varying matrix of appropriate dimensions.

The functions $f(k, x(k), \xi(k)) \in \mathbb{R}^{n_x}$ and $g_i(k, x(k), \zeta_i(k)) \in \mathbb{R}^{n_y}$ represent the stochastic nonlinearities satisfying $f(k, 0, \xi(k)) = 0$, $g_i(k, 0, \zeta_i(k)) = 0$ and the following statistics:

$$\begin{aligned} &\mathbb{E}\{f(k, x(k), \xi(k))|x(k)\} = 0,\\ &\mathbb{E}\{g_i(k, x(k), \zeta_i(k))|x(k)\} = 0, \end{aligned} \tag{8.3}$$

$$\mathbb{E}\left\{g_i(k, x(k), \zeta_i(k))g_j'(k, x(k), \zeta_j(k))|x(k)\right\} = 0,\ i \neq j,$$

$$\mathbb{E}\left\{\begin{bmatrix} f(k, x(k), \xi(k)) \\ g_i(k, x(k), \zeta_i(k)) \end{bmatrix}\begin{bmatrix} f(s, x(s), \xi(s)) \\ g_i(s, x(s), \zeta_i(s)) \end{bmatrix}'\middle| x(k)\right\} = 0,$$

$$k \neq s,$$

$$\begin{aligned}&\mathbb{E}\left\{\begin{bmatrix} f(k,x(k),\xi(k)) \\ g_i(k,x(k),\zeta_i(k)) \end{bmatrix}\begin{bmatrix} f(k,x(k),\xi(k)) \\ g_i(k,x(k),\zeta_i(k)) \end{bmatrix}'\Bigg| x(k)\right\} \\ &\qquad = \sum_{s=1}^{m} \Pi_s(k)x'(k)\Gamma_s(k)x(k),\end{aligned} \tag{8.4}$$

where m is a given positive integer, and $\Pi_s(k) = \text{diag}\left\{\Pi_{s1}(k), \Pi_{s2}^{(i)}(k)\right\}$. $\Pi_{s1}(k)$, $\Pi_{s2}^{(i)}(k)$, and $\Gamma_s(k)$ are known matrices with compatible dimensions for $s = 1, 2, \ldots, m$.

Remark 8.1 As has been discussed in [13], the stochastic nonlinearities $f(k, x(k), \xi(k))$ and $g_i(k, x(k), \zeta_i(k))$ introduced in this chapter are quite general that cover some commonly investigated nonlinearities as special cases such as state-multiplicative noises (i.e., $A(k)x(k)\xi(k)$) and random vectors whose power relies on the sign of the state (i.e., $\text{sgn}(x(k))A(k)x(k)\xi(k)$).

8.1.2 Distributed Resilient Filter

A fundamental issue in wireless sensor networks is to design the filters so as to restore the state vector in a cooperative behavior. Note that, in practical applications, gain variations often occur in the implementation of a filter due probably to computational or tuning uncertainties. Since the performance of the filter may be susceptible to the perturbations in gain parameters, the design of resilient filters capable of tolerating some level of gain variations is of engineering significance.

To observe the target plant through a network of interconnected sensors, a two-step distributed estimator is proposed as follows:

$$\hat{x}_i(k|k-1) = A(k-1)\hat{x}_i(k-1|k-1), \tag{8.5a}$$

$$\hat{x}_i(k|k) = \hat{x}_i(k|k-1) + \sum_{j\in N_i} a_{ij}(G_{ij}(k) + \Delta_{ij}(k))\tilde{y}_j(k), \tag{8.5b}$$

with the initial value $\hat{x}_i(0|0) = \mathbb{E}[x(0)] = \mu_0$, for $i \in V$. Note that $\hat{x}_i(k|k-1)$ and $\hat{x}_i(k|k)$ are the one-step prediction and the estimate of state vector $x(k)$, respectively. $z\tilde{y}_i(k) = y_i(k) - \bar{\gamma}_i(k)C_i(k)\hat{x}_i(k|k-1)$ is the innovation sequence exchanged via the network. The matrix $G_{ij}(k) \in \mathbb{R}^{n_x\times n_y}$ represents the gain coefficients of the filters to be designed. The term $\Delta_{ij}(k) \in \mathbb{R}^{n_x\times n_y}$ models the computational or implementation error associated with the estimator gain, and is assumed to have zero mean and a bounded second moment, i.e.,

$$E[\Delta_{ij}(k)] = 0, \quad E[\Delta_{ij}(k)\Delta'_{ij}(k)] \leq \delta_{ij} I, \tag{8.6}$$

where δ_{ij} is a positive scalar. Moreover, throughout the chapter, we assume that all the stochastic variables, i.e., $\Delta_{ij}(k)$, $\gamma_i(k)$, $\xi(k)$, $\zeta_i(k)$, $x(0)$, $w(k)$, and $v_i(k)$, are white and mutually independent.

For the convenience of later development, let us define the local state prediction and local state estimation error vectors, respectively, as follows:

$$\begin{aligned} e_i(k|k-1) &\triangleq x(k) - \hat{x}_i(k|k-1) \\ e_i(k|k) &\triangleq x(k) - \hat{x}_i(k|k). \end{aligned}$$

Substituting (8.5a) into the state prediction error equation yields

$$e_i(k+1|k) = A(k)e_i(k|k) + f(k, x(k), \xi(k)) + w(k), \tag{8.7}$$

and it can then be seen from (8.5b) that the dynamics of the estimation errors evolves according to

$$\begin{aligned} e_i(k|k) =& e_i(k|k-1) - \sum_{j \in N_i} a_{ij}(G_{ij}(k) + \Delta_{ij}(k))\Big\{v_j(k) \\ &+ g_j(k, x(k), \zeta_j(k)) + \tilde{\gamma}_j(k)C_j(k)x(k) \\ &+ \bar{\gamma}_j(k)C_j(k)e_j(k|k-1)\Big\}, \end{aligned} \tag{8.8}$$

where $\tilde{\gamma}_i(k) = \gamma_i(k) - \bar{\gamma}_i(k)$. For the sake of simplicity, we denote

$$\begin{aligned} e(k|k-1) &\triangleq \mathrm{col}_n\{e_i(k|k-1)\}, \\ e(k|k) &\triangleq \mathrm{col}_n\{e_i(k|k)\}, \end{aligned}$$

and then (8.7)–(8.8) can be rearranged into a more compact form as follows:

$$\begin{aligned} e(k+1|k) =& \mathcal{A}(k)e(k|k) + \vec{f}(k) + \vec{w}(k), \\ e(k|k) =& e(k|k-1) - \sum_{i=1}^{n} E_i(G(k) + \Delta(k))H_i\Big\{\vec{g}(k) + \vec{v}(k) \\ &+ \tilde{\Gamma}(k)\mathcal{C}(k)\vec{x}(k) + \bar{\Gamma}(k)\mathcal{C}(k)e(k|k-1)\Big\}, \end{aligned}$$

where

$$
\begin{aligned}
&\mathcal{A}(k) \triangleq \mathrm{diag}_n\{A(k)\},\ \vec{f}(k) \triangleq \mathrm{col}_n\{f(k, x(k), \xi(k))\},\\
&\mathcal{C}(k) \triangleq \mathrm{diag}_n\{C_i(k)\},\ \vec{g}(k) \triangleq \mathrm{col}_n\{g_i(k, x(k), \zeta_i(k))\}\\
&\vec{w}(k) \triangleq \mathrm{col}_n(w(k)),\ H_i = \mathrm{diag}\{a_{i1}I, \ldots, a_{in}I\}\\
&\Delta(k) \triangleq \{\Delta_{ij}(k)\}_{n\times n},\ G(k) \triangleq \{G_{ij}(k)\}_{n\times n}\\
&\vec{v}(k) \triangleq \mathrm{col}_n\{v_i(k)\},\ \tilde{\Gamma}(k) \triangleq \mathrm{diag}_n\{\tilde{\gamma}_i(k)I\},\\
&\vec{x}(k) \triangleq \mathrm{col}_n(x(k)),\ \bar{\Gamma}(k) \triangleq \mathrm{diag}_n\{\bar{\gamma}_i(k)I\}\\
&E_i \triangleq \mathrm{diag}\{\underbrace{0, \ldots, 0}_{i-1}, I, \underbrace{0, \ldots, 0}_{n-i}\}.
\end{aligned}
$$

Furthermore, by letting $K(k) = -\sum_{i=1}^{n} E_i(G(k) + \Delta(k))H_i$, we have

$$
\begin{aligned}
e(k|k) =&(I + K(k)\bar{\Gamma}(k)\mathcal{C}(k))e(k|k-1)\\
&+ K(k)\Big\{\vec{g}(k) + \vec{v}(k) + \tilde{\Gamma}(k)\mathcal{C}(k)\vec{x}(k)\Big\}.
\end{aligned}
\tag{8.9}
$$

To quantify the transient performance of the proposed distributed resilient filter, a finite horizon quadratic filtering cost function is introduced for the wireless sensor networks as follows:

$$
J_T(\mathbf{G}(T)) = \sum_{k=0}^{T}\sum_{i=1}^{n} \mathbb{E}[e_i'(k|k)e_i(k|k)],
\tag{8.10}
$$

where the set $\mathbf{G}(T) = \{G(k),\ k = 1, 2, \ldots, T-1\}$ gathers the filter coefficients in all the T steps. Define the error covariances as $P_{k|k-1} \triangleq \mathbb{E}[e(k|k-1)e'(k|k-1)]$ and $P_{k|k} \triangleq \mathbb{E}[e(k|k)e'(k|k)]$. Obviously, the above quadratic filtering cost function can be rewritten as $J_T(\mathbf{G}(T)) = \sum_{k=0}^{T} \mathrm{tr}\left\{P_{k|k}\right\}$. In the chapter, we aim to design the optimal distributed filters by solving the following optimization problem:

$$
J_T = \arg\min_{G(T-1)} J_T(\mathbf{G}(T)).
\tag{8.11}
$$

8.2 Preliminary

In this section, some preliminary knowledge is derived for preparation. At the very beginning, the following lemmas are introduced, which will be used to establish our main results.

Lemma 8.2.1 ([24]) *Let* $D = [d_{ij}]_{p\times p}$ *be a real-valued matrix and* $B = \text{diag}\{b_1, b_2, \ldots, b_p\}$ *be a diagonal random matrix. Then*

$$\mathbb{E}[BDB'] = \begin{pmatrix} \mathbb{E}[b_1^2] & \mathbb{E}[b_1 b_2] & \cdots & \mathbb{E}[b_1 b_p] \\ \mathbb{E}[b_2 b_1] & \mathbb{E}[b_2^2] & \cdots & \mathbb{E}[b_2 b_p] \\ \vdots & \vdots & \ddots & \vdots \\ \mathbb{E}[b_p b_1] & \mathbb{E}[b_p b_2] & \cdots & \mathbb{E}[b_p^2] \end{pmatrix} \odot D,$$

where $\odot$ *is the Hadamard product.*

Lemma 8.2.2 *Consider a discrete time-varying process with stochastic nonlinearities given by* (8.1). *The system state covariance* $X(k) \triangleq E[x(k)x'(k)]$ *satisfies the following recursive equation:*

$$\begin{aligned} X(k+1) =& A(k)X(k)A'(k) \\ &+ \sum_{s=1}^{m} \Pi_{s1}(k)\text{tr}\,\{X(k)\Gamma_s(k)\} + Q(k). \end{aligned}$$

Proof Substituting (8.1) into $X(k+1)$ yields

$$\begin{aligned} &X(k+1) \\ &= A(k)X(k)A'(k) + \mathbb{E}[f(k)f'(k)] + \mathbb{E}[w(k)w'(k)], \end{aligned}$$

where the cross terms vanish because $x(k)$, $f(k)$, and $w(k)$ are mutually independent. From (8.4), it can be seen that

$$\mathbb{E}[f(k)f'(k)] = \sum_{s=1}^{m} \Pi_{s1}(k)\mathbb{E}[x'(k)\Gamma_s(k)x(k)].$$

Note that $\mathbb{E}[x'(k)\Gamma_s(k)x(k)]$ is a scalar, so its value will not be changed by taking its trace as follows:

$$\mathbb{E}[x'(k)\Gamma_s(k)x(k)] = \text{tr}\,\{X(k)\Gamma_s(k)\},$$

which concludes the proof.

The following lemma gives the dynamic evolution of the prediction error covariance.

Lemma 8.2.3 *Given the error covariance* $P_{k|k} > 0$ *at step* k, *the prediction error covariance satisfies the following dynamic equation:*

$$P_{k+1|k} = \mathcal{A}(k)P_{k|k}\mathcal{A}'(k) + \mathcal{F}(k) + \mathcal{W}(k), \tag{8.12}$$

where

$$\mathcal{F}(k) = \mathbf{1}_n \otimes \sum_{s=1}^{m} \Pi_{s1}(k)\mathrm{tr}\left\{X(k)\Gamma_s(k)\right\},$$

$$\mathcal{W}(k) = \mathbf{1}_n \otimes Q(k).$$

Here, $\mathbf{1}_n \in \mathbb{R}^{n\times n}$ *is a square matrix with all the elements equal to one and* $\otimes$ *is the Kronecker product.*

Proof The proof of this lemma is similar to Lemma 8.2.2, and thus is omitted here for sake of brevity.

Lemma 8.2.4 *The two-step distributed filters given by* (8.5a)–(8.5b) *are unbiased and the corresponding estimation error covariance can be given as follows:*

$$\begin{aligned} P_{k|k} = {} & \mathbb{E}[(I + K(k)\bar{\Gamma}(k)\mathcal{C}(k))P_{k|k-1}(I + K(k)\bar{\Gamma}(k)\mathcal{C}(k))'] \\ & + \mathbb{E}[K(k)\vec{g}(k)\vec{g}'(k)K'(k)] + \mathbb{E}[K(k)\vec{x}(k)\vec{x}'(k)K(k)] \\ & + \mathbb{E}[K(k)\tilde{\Gamma}(k)\mathcal{C}(k)\vec{x}(k)\vec{x}'(k)\mathcal{C}'(k)\tilde{\Gamma}'(k)K'(k)]. \end{aligned} \tag{8.13}$$

Proof In view of the initial estimate $\hat{x}_i(0|0) = \mathbb{E}\{x(0)\}$, it is obvious that $\mathbb{E}\{e(0|0)\} = 0$. Using the fact that the stochastic nonlinearities and measurement noises are of zero means, we obtain $\mathbb{E}\{e(1|0)\} = 0$. Furthermore, we have $\mathbb{E}\{e(1|1)\} = 0$ since $\mathbb{E}[\tilde{\gamma}_i(k)] = 0$. By repeating such a procedure, it can be concluded that $\mathbb{E}[e(k|k-1)] = 0$, $\mathbb{E}[e(k|k)] = 0$. Thus, the unbiasedness of the proposed distributed filters can be guaranteed. As for the error covariance, by applying (8.9), we arrive at the following equation:

$$\begin{aligned} P_{k|k} = {} & \mathbb{E}\left[(I + K(k)\bar{\Gamma}(k)\mathcal{C}(k))P_{k|k-1}(I + K(k)\bar{\Gamma}(k)\mathcal{C}(k))'\right] \\ & + \mathcal{P} + \mathcal{P}' + \mathbb{E}\left[K(k)\vec{g}(k)\vec{g}'(k)K'(k)\right] + \mathcal{L} + \mathcal{L}' \\ & + \mathbb{E}\left[K(k)\vec{v}(k)\vec{v}'(k)K(k)\right] + \mathcal{R} + \mathcal{R}' \\ & + \mathbb{E}\left[K(k)\tilde{\Gamma}(k)\mathcal{C}(k)\vec{x}(k)\vec{x}'(k)\mathcal{C}'(k)\tilde{\Gamma}'(k)K'(k)\right], \end{aligned}$$

where

$$\begin{aligned} \mathcal{P} = {} & \mathbb{E}[(I + K(k)\bar{\Gamma}(k)\mathcal{C}(k))e(k|k-1)(\vec{g}(k) + \vec{v}(k), \\ & + \tilde{\Gamma}(k)\mathcal{C}(k)\vec{x}(k))'K'(k)], \\ \mathcal{L} = {} & \mathbb{E}[K(k)\vec{g}(k)\vec{v}'(k)K'(k)] + \mathbb{E}[K(k)\vec{g}(k)\vec{x}'(k)\mathcal{C}'(k)\tilde{\Gamma}'(k)] \\ \mathcal{R} = {} & \mathbb{E}[K(k)\vec{v}(k)\vec{x}'(k)\mathcal{C}'(k)\tilde{\Gamma}'(k)K'(k)]. \end{aligned}$$

Noting that the prediction error vector $e(k|k-1)$ is uncorrelated with $\vec{g}(k)$, $\vec{v}(k)$, and $\tilde{\Gamma}(k)$, we have the term $\mathcal{P}$ vanished. Also, exploiting the fact that the noises $\vec{g}(k)$

and $\vec{v}(k)$ are independent with each other and the initial state is $x(0)$, one can derive that $\mathcal{L} = 0$ and $\mathcal{R} = 0$, which ends the proof.

In the above lemma, the dynamics of the error covariance is presented, which turns out to be dependent on determined by the network topology, the state transition matrix, the measurement matrices and the statistics of stochastic nonlinearities, sensor gain degradations, and external disturbances. Unfortunately, due primarily to the existence of gain variations in this chapter, it is impossible to solve the error covariance analytically, not to mention the further design of the optimal gain coefficients. An alternative yet effective way for designing the filters is to establish an upper bound on the estimation error covariance.

Suppose that, for all $k \geq 0$, there exist positive definite matrices $M_{k|k} \geq P_{k|k}$. Define a finite horizon quadratic cost function as follows:

$$\bar{J}_T(\mathbf{G}(T)) = \sum_{k=0}^{T} \operatorname{tr}\left\{M_{k|k}\right\}.$$

It is clear that $J_T(\mathbf{G}(T)) \leq \bar{J}_T(\mathbf{G}(T))$ which implies that $\bar{J}_T(\mathbf{G}(T))$ is an upper bound on the original cost function. As a result, we can focus on minimizing this upper bound by appropriately choosing the filter parameters, namely,

$$\bar{J}_T = \arg \min_{G(T-1)} \bar{J}_T(\mathbf{G}(T)).$$

A distinguished advantage of the above proposed scheme lies in that it can solve some sort of optimization problems where the analytical expression of the objective function is unavailable. By constructing an analytical upper bound, one can provide an alternative, feasible, yet suboptimal option for the filter design under guaranteed performance.

8.3 Suboptimal Distributed Resilient Filter Design

In this section, let us first derive an analytical upper bound for the estimation error covariance of the system (8.9), and then design the gain $G(k)$ of the suboptimal filter in order to minimize the upper bound at each time-step k.

For presentation convenience, some notations are introduced as follows:

$$\begin{aligned}
\bar{K}(k) &\triangleq -\sum_{i=1}^{n} E_i G(k) H_i, \quad \tilde{K}(k) \triangleq -\sum_{i=1}^{n} E_i \Delta(k) H_i, \\
\Upsilon(k) &\triangleq \vec{g}(k)\vec{g}'(k) + \vec{v}(k)\vec{v}'(k) \\
&\quad + \tilde{\Gamma}(k)\mathcal{C}(k)\vec{x}(k)\vec{x}'(k)\mathcal{C}'(k)\tilde{\Gamma}'(k).
\end{aligned}$$

Moreover, from the definition of the stochastic nonlinearity $g_i(k, x(k), \zeta_i(k))$, we have

$$\aleph(k) \triangleq \mathbb{E}[\vec{g}(k)\vec{g}'(k)] = \sum_{s=1}^{m} \Pi_{s2}(k)\text{tr}[X(k)\Gamma_s(k)],$$

where $\Pi_{s2}(k) \triangleq \text{diag}\left\{\Pi_{s2}^{(1)}(k), \ldots, \Pi_{s2}^{(n)}(k)\right\}$. Additionally, in light of Lemma 8.2.1, we have

$$\mathbb{E}\left[\tilde{\Gamma}(k)\mathcal{C}(k)\vec{x}(k)\vec{x}'(k)\mathcal{C}'(k)\tilde{\Gamma}'(k)\right] = \Xi(k) \odot \left(\mathcal{C}(k)\mathcal{X}(k)\mathcal{C}'(k)\right),$$

where $\mathcal{X}(k) \triangleq \mathbf{1}_n \otimes X(k)$ and $\Xi(k) \triangleq \text{diag}\{\sigma_{\gamma_1}^2(k)\mathbf{1}_n, \ldots, \sigma_{\gamma_n}^2(k)\mathbf{1}_n\}$. Therefore, the mean value $\bar{\Upsilon}(k) \triangleq \mathbb{E}[\Upsilon(k)]$ can be computed by

$$\bar{\Upsilon}(k) = \aleph(k) + \text{diag}_n\{R_i(k)\} + \Xi(k) \odot \left(\mathcal{C}(k)\mathcal{X}(k)\mathcal{C}'(k)\right).$$

Now, we can derive the upper bound of $P_{k|k}$ in the following theorem.

Theorem 8.3.1 *Consider the following difference equations:*

$$M_{k+1|k} = \mathcal{A}(k)M_{k|k}\mathcal{A}'(k) + \mathcal{F}(k) + \mathcal{W}(k), \tag{8.14a}$$

$$\begin{aligned} M_{k|k} &= (I + \bar{K}(k)\bar{\Gamma}(k)\mathcal{C}(k))M_{k|k-1}(I + \bar{K}(k)\bar{\Gamma}(k)\mathcal{C}(k))' \\ &\quad + \lambda_{max}(\bar{\Gamma}(k)\mathcal{C}(k)M_{k|k-1}\mathcal{C}'(k)\bar{\Gamma}(k) + \bar{\Upsilon}(k))\Lambda \\ &\quad + \bar{K}(k)\bar{\Upsilon}(k)\bar{K}'(k) \end{aligned} \tag{8.14b}$$

with the initial condition $M_{0|0} = P_{0|0} = \Sigma_0$, where $\Lambda \triangleq \text{diag}\left\{\sum_{s=1}^{n} a_{1s}^2\delta_{1s}I, \ldots, \sum_{s=1}^{n} a_{ns}^2\delta_{ns}I\right\}$. Then, the inequalities $P_{k|k} \leq M_{k|k}$ and $P_{k+1|k} \leq M_{k+1|k}$ always hold for all $k \geq 0$.

Proof Since the uncertainty $\Delta(k)$ is of zero mean and independent with other stochastic variables, (8.13) can be rewritten in the following form:

$$\begin{aligned} P_{k|k} &= (I + \bar{K}(k)\bar{\Gamma}(k)\mathcal{C}(k))P_{k|k-1}(I + \bar{K}(k)\bar{\Gamma}(k)\mathcal{C}(k))' \\ &\quad + \mathbb{E}[\tilde{K}(k)\bar{\Gamma}(k)\mathcal{C}(k)P_{k|k-1}\mathcal{C}'(k)\bar{\Gamma}(k)'\tilde{K}'(k)] \\ &\quad + \mathbb{E}[\tilde{K}(k)\Upsilon(k)\tilde{K}'(k)] + \bar{K}(k)\bar{\Upsilon}(k)\bar{K}'(k). \end{aligned} \tag{8.15}$$

Subsequently, let us prove this theorem by induction. Assume, inductively, that $P_{k-1|k-1} \leq M_{k-1|k-1}$. Applying (8.12) and (8.14a), we have

$$
\begin{aligned}
&P_{k|k-1} - M_{k|k-1} \\
&= \mathcal{A}(k-1)(P_{k-1|k-1} - M_{k-1|k-1})\mathcal{A}'(k-1) \leq 0,
\end{aligned}
$$

which implies $P_{k|k-1} \leq M_{k|k-1}$. The difference $P_{k|k} - M_{k|k}$ can be written as

$$
\begin{aligned}
&P_{k|k} - M_{k|k} \\
&\leq \mathbb{E}\Big[\tilde{K}(k)(\bar{\Gamma}(k)\mathcal{C}(k)P_{k|k-1}\mathcal{C}'(k)\bar{\Gamma}(k)' + \Upsilon(k))\tilde{K}'(k)\Big] \\
&\quad - \lambda_{max}\Big(\bar{\Gamma}(k)\mathcal{C}(k)M_{k|k-1}\mathcal{C}'(k)\bar{\Gamma}(k) + \bar{\Upsilon}(k)\Big)\Lambda.
\end{aligned} \tag{8.16}
$$

Moreover, since $P_{k|k-1} \leq M_{k|k-1}$, it follows that

$$
\begin{aligned}
&\mathbb{E}\Big[\tilde{K}(k)(\bar{\Gamma}(k)\mathcal{C}(k)P_{k|k-1}\mathcal{C}'(k)\bar{\Gamma}(k)' + \Upsilon(k))\tilde{K}'(k)\Big] \\
&\leq \lambda_{\max}\Big(\bar{\Gamma}(k)\mathcal{C}(k)M_{k|k-1}\mathcal{C}'(k)\bar{\Gamma}(k) + \bar{\Upsilon}(k)\Big)\mathbb{E}\Big[\tilde{K}(k)\tilde{K}'(k)\Big].
\end{aligned}
$$

Now, we are in the position to tackle the term in the right-hand side of the above equation. Utilizing algebraic transformations, it is not difficult to verify that

$$
\begin{aligned}
\tilde{K}(k)\tilde{K}'(k) &= \left(\sum_{i=1}^{n} E_i\Delta(k)H_i\right)\left(\sum_{i=1}^{n} E_i\Delta(k)H_i\right)' \\
&= \left\{\left(\sum_{s=1}^{n} a_{is}a_{js}\Delta_{is}(k)\Delta'_{js}(k)\right)_{ij}\right\}_{n\times n}.
\end{aligned}
$$

Based on (8.6) and the fact that $\Delta_{ij}(k)$ are mutually independent for $i, j = 1, \ldots, n$, one has

$$
\mathbb{E}\Big[\tilde{K}(k)\tilde{K}'(k)\Big] \leq \Lambda. \tag{8.17}
$$

Together with (8.16)–(8.17), we can see $P_{k|k} \leq M_{k|k}$. The inductive hypothesis implies that $P_{k|k} \leq M_{k|k}$, which completes the proof.

In the next step, we will design the optimal filter gains such that the upper bound $P_{k|k}$ can be minimized at each step. Before proceeding further, let us define $G^{(i)}(k)$ to be the ith row of the block matrix $G(k)$, i.e.,

$$
G^{(i)}(k) \triangleq \big[G_{i1}(k), \ldots, G_{in}(k)\big]
$$

and $M^{(i)}_{k|k-1}$ to be the ith row of the block matrix $M_{k|k-1}$. Moreover, define

$$\mathcal{M}_i(k) \triangleq H_i\Big[\bar{\Gamma}(k)\mathcal{C}(k)M_{k|k-1}\mathcal{C}'(k)\bar{\Gamma}(k) + \bar{\Upsilon}(k)\Big]H_i,$$
$$\mathcal{N}_i(k) \triangleq M^{(i)}_{k|k-1}\mathcal{C}'(k)\bar{\Gamma}(k)H_i.$$

By removing the b-th ($b \notin N_i$) column block from the matrices $\mathcal{N}_i(k)$ and $G^{(i)}(k)$, one can obtain $\bar{\mathcal{N}}_i(k)$ and $\bar{G}^{(i)}(k)$, respectively. In addition, we let $\bar{\mathcal{M}}_i(k)$ be a simplified matrix by removing both the b-th row and b-th column block from $\mathcal{M}_i(k)$, when $b \notin N_i$.

Theorem 8.3.2 *Consider the time-varying system* (8.1)–(8.2) *with distributed resilient filters given by* (8.5a)–(8.5b)*. The upper bound of the error covariance* (8.14a)–(8.14b) *can be minimized at each step by choosing the parameters of filters as follows:*

$$G_{ij}(k) = \begin{cases} 0 & a_{ij} = 0 \\ \left(\bar{\mathcal{N}}_i(k)\bar{\mathcal{M}}_i(k)^{-1}\right)^{\sharp}, & a_{ij} \neq 0 \end{cases} \tag{8.18}$$

where $(*)^{\sharp}$ *extracts the corresponding submatrix from the matrix "$*$" associated with the parameter* $G_{ij}(k)$.

Proof Taking the trace for the both sides of (8.14b) yields that

$$\begin{aligned}
&\mathrm{tr}\{M_{k|k}\} \\
&= \mathrm{tr}\Big\{(I + \bar{K}(k)\bar{\Gamma}(k)\mathcal{C}(k))M_{k|k-1}(I + \bar{K}(k)\bar{\Gamma}(k)\mathcal{C}(k))'\Big\} \\
&+ \mathrm{tr}\big\{\bar{K}(k)\bar{\Upsilon}(k)\bar{K}'(k)\big\} + \sum_{i=1}^{n}\sum_{j=1}^{n} a_{ij}^2\delta_{ij}n_x \\
&\times \lambda_{max}\Big(\bar{\Gamma}(k)\mathcal{C}(k)M_{k|k-1}\mathcal{C}'(k)\bar{\Gamma}(k) + \bar{\Upsilon}(k)\Big).
\end{aligned} \tag{8.19}$$

The first term in the right-hand side of (8.19) can be rewritten into the following expression:

$$\begin{aligned}
&\mathrm{tr}\Big\{(I + \bar{K}(k)\bar{\Gamma}(k)\mathcal{C}(k))M_{k|k-1}(I + \bar{K}(k)\bar{\Gamma}(k)\mathcal{C}(k))'\Big\} \\
&= \mathrm{tr}\{M_{k|k-1}\} + 2\mathrm{tr}\{\bar{K}(k)\bar{\Gamma}(k)\mathcal{C}(k)M_{k|k-1}\} \\
&\quad + \mathrm{tr}\{\bar{K}(k)\bar{\Gamma}(k)\mathcal{C}(k)M_{k|k-1}\mathcal{C}'(k)\bar{\Gamma}(k)'\bar{K}'(k)\}.
\end{aligned}$$

Resorting to the properties of trace, we have

$$\begin{aligned}
&\mathrm{tr}\{E_iG(k)H_iMH_iG'(k)E_j\} \\
&= \mathrm{tr}\{E_jE_iG(k)H_iMH_iG'(k)\} = 0, \quad \text{for } i \neq j
\end{aligned} \tag{8.20}$$

for an arbitrary matrix M with appropriate dimensions. Noticing the definition of $\bar{K}(k)$ and exploiting (8.20), it is obvious that

$$\begin{aligned}
&\mathrm{tr}\big\{\bar{K}(k)\bar{\Gamma}(k)\mathcal{C}(k)M_{k|k-1}\mathcal{C}'(k)\bar{\Gamma}(k)\bar{K}'(k)\big\} \\
&= \mathrm{tr}\left\{\sum_{i=1}^{n} E_i G(k) H_i \bar{\Gamma}(k)\mathcal{C}(k)M_{k|k-1}\mathcal{C}'(k)\bar{\Gamma}(k)H_i G'(k)\right\}.
\end{aligned}$$

As for the second term in the right-hand side of (8.19), one can derive that

$$\begin{aligned}
&\mathrm{tr}\{\bar{K}(k)\bar{\Upsilon}(k)\bar{K}'(k)\} \\
&\quad = \mathrm{tr}\left\{\sum_{i=1}^{n} E_i G(k) H_i \bar{\Upsilon}(k) H_i G'(k)\right\}.
\end{aligned}$$

Moreover, taking the partial derivation of the trace of the matrix $M_{k|k}$ with respect to the gain parameters $G(k)$, we have

$$\begin{aligned}
&\frac{\partial \mathrm{tr}\{M_{k|k}\}}{\partial G(k)} \\
&= -2\sum_{i=1}^{n} E_i M_{k|k-1}\mathcal{C}'(k)\bar{\Gamma}(k)H_i + 2\sum_{i=1}^{n} E_i G(k) H_i \\
&\quad \times \Big[\bar{\Gamma}(k)\mathcal{C}(k)M_{k|k-1}\mathcal{C}'(k)\bar{\Gamma}(k) + \bar{\Upsilon}(k)\Big]H_i.
\end{aligned}$$

Since (8.19) is in a positive semi-definite quadratic form with respect to the matrix $G(k)$, in order to minimize $\mathrm{tr}\{M_{k|k}\}$, we let its partial derivative be zero. As such, we have

$$\begin{aligned}
&\sum_{i=1}^{n} E_i G(k) H_i \Big[\bar{\Gamma}(k)\mathcal{C}(k)M_{k|k-1}\mathcal{C}'(k)\bar{\Gamma}(k) + \bar{\Upsilon}(k)\Big]H_i \\
&= \sum_{i=1}^{n} E_i M_{k|k-1}\mathcal{C}'(k)\bar{\Gamma}(k)H_i, \qquad (8.21)
\end{aligned}$$

which is equal to the following equations containing sparse matrices:

$$G^{(i)}(k)\mathcal{M}_i(k) = \mathcal{N}_i(k)$$

for $i = 1, \ldots, n$. Now, it remains to obtain $G^{(i)}(k)$ by solving the above equations. Note that

$$H_i = \mathrm{diag}\{a_{i1}I,\ a_{i2}I, \ldots,\ a_{in}I\},$$

where the element $a_{ij} = 0$ when $j \notin N_i$. Removing the corresponding zero rows and zero columns from $\mathcal{M}_i(k)$ and the corresponding zero columns from $\mathcal{N}_i(k)$, we obtain the simplified matrices $\bar{\mathcal{M}}_i(k)$ and $\bar{\mathcal{N}}_i(k)$. Subsequently, we can partition the matrices by $\mathcal{M}_i(k) = \{M_{ab}(k)\}_{n\times n}$ and $\mathcal{N}_i(k) = \{N_b(k)\}_{1\times n}$, and thus have

$$\sum_{j=1}^{n} G_{ij}(k)M_{jb}(k) = N_b(k), \quad \text{for } b \in N_i.$$

As for $b \notin N_i$, we have $M_{jb}(k) = N_b(k) = 0$, and therefore, the above equation always holds. Moreover, for $b \notin N_i$, we can choose the coupling filter gains $G_{ib}(k) = 0$ because the local sensor cannot receive any message from its non-neighbor nodes. As a result, it follows that

$$\bar{G}^{(i)}(k)\bar{\mathcal{M}}_i(k) = \bar{\mathcal{N}}_i(k), \quad \text{for } i = 1, 2, \ldots, n.$$

Noticing that the matrix $\bar{\mathcal{M}}_i(k)$ is positive definite, we derive that $\bar{G}^{(i)}(k) = \bar{\mathcal{N}}_i(k)\bar{\mathcal{M}}_i(k)^{-1}$ and, consequently, the parameter $G_{ij}(k)$ can be obtained by selecting the corresponding column block matrix in the matrix $\bar{\mathcal{N}}_i(k)\bar{\mathcal{M}}_i(k)^{-1}$, which ends the proof.

Remark 8.2 A crucial step for designing the filter gain is to solve the equality (8.21). However, due to the sparsity of the communication topology, there is a remarkable difficulty to obtain $G(k)$ directly. Actually, the diagonal entries of matrix H_i are nonzero when the corresponding sensor is in the neighboring set of sensor i. In other words, $a_{ij} > 0$ if only if $j \in N_i$, and therefore H_i is likely to be rank deficient, which means that $\mathcal{M}_i(k)$ is also rank deficient. By employing the matrix simplification technique proposed in the above proof, we remove the zero columns and rows to guarantee the positive definiteness of the simplified matrix $\bar{\mathcal{M}}_i(k)$, which renders the explicit expression of $G(k)$ possible.

8.4 Boundedness Analysis

In this section, we will discuss the mean-square boundedness of the estimation errors for the proposed distributed resilient filter.

For convenience of discussion, without loss of generality, we set the weights $a_{ij} = 1$ for $j \in N_i$. Moreover, an assumption is introduced to place some constraints on the system parameters.

Assumption 8.1 There exist positive real numbers $\bar{a}$, $\underline{c}$, $\bar{c}$, $\bar{\tau}$, $\underline{\lambda}$, $\bar{\lambda}$, $\bar{q}$, $\bar{r}$, κ_{si}, κ_s, $\bar{\sigma}^2$, such that the following bounds on matrices are fulfilled for all $i = 1, \ldots, n$, $j = 1, 2$, and $s = 1, \ldots, m$:

$$\|A(k)\| \le \bar{a},\ \underline{c} \le \|C_i(k)\|_{\min},\ \|C_i(k)\| \le \bar{c},\ \mathrm{tr}\{X(k)\} \le \bar{\tau},$$
$$\|Q(k)\| \le \bar{q},\ \underline{\lambda} \le \|\bar{\Gamma}(k)\|_{\min},\ \|\bar{\Gamma}(k)\| \le \bar{\lambda},\ \|R_i(k)\| \le \bar{r},$$
$$\|\Pi_{sj}(k)\| \le \kappa_{sj},\ \|\Gamma_s(k)\| \le \kappa_s,\ \sigma_{\gamma_i}^2 \le \bar{\sigma}^2.$$

Denote $\bar{k} \triangleq n\bar{\lambda}\bar{c}/\underline{\lambda}^2\underline{c}^2$ and $\bar{\zeta} \triangleq \max_i \left\{\sum_{s=1}^n \delta_{is}\right\}$. With Assumption 8.1, we are able to establish a sufficient condition for the mean-square boundedness of the estimation errors as follows.

Theorem 8.4.1 *Consider the time-varying system* (8.1)–(8.2) *with the distributed resilient filters given by* (8.5a)–(8.5b) *whose gain parameters are provided in Theorem 8.3.2. Under Assumption 8.1, the filtering error dynamics is mean-square bounded, i.e.,*

$$\sup_{k\in\mathbb{N}} \sum_{i=1}^n \mathbb{E}[e_i'(k|k)e_i(k|k)] < \infty,$$

if the following inequality holds

$$\bar{a}^2((1+\bar{k}\bar{\lambda}\bar{c})^2 + \bar{\lambda}^2\bar{c}^2\bar{\zeta}) < 1. \tag{8.22}$$

Proof It follows from (8.14a) and Assumption 8.1 that

$$\|M_{k|k-1}\| \le \bar{a}^2\|M_{k-1|k-1}\| + \|\mathcal{F}(k)\| + \|\mathcal{W}(k)\|.$$

Noting that

$$\mathrm{tr}\{X(k)\Gamma_s(k)\} = \mathrm{tr}\left\{\mathbb{E}[x'(k)\Gamma_s(k)x(k)]\right\} \le \kappa_s\bar{\tau},$$

one has

$$\|\mathcal{F}(k)\| \le n\sum_{s=1}^m \kappa_{s1}\kappa_s\bar{\tau}.$$

In addition, it can be seen that $\|\mathcal{W}(k)\| \le n\bar{q}$. Therefore, we can obtain the following inequality:

$$\|M_{k|k-1}\| \le \bar{a}^2\|M_{k-1|k-1}\| + n\sum_{s=1}^m \kappa_{s1}\kappa_s\bar{\tau} + n\bar{q}. \tag{8.23}$$

Since we only care about the non-sparse part of $G(k)$, it is not difficult to verify that (8.21) results in the following equation:

$$\sum_{i=1}^n E_iG(k)H_i = \sum_{i=1}^n E_iU(k)(Z(k))^{-1}H_i, \tag{8.24}$$

where

$$U(k) \triangleq M_{k|k-1}\mathcal{C}'(k)\bar{\Gamma}(k),$$
$$Z(k) \triangleq \bar{\Gamma}(k)\mathcal{C}(k)M_{k|k-1}\mathcal{C}'(k)\bar{\Gamma}(k) + \bar{\Upsilon}(k).$$

Taking the norm for the both sides of the equation (8.24) yields that

$$\|\bar{K}(k)\| \leq n\|U(k)(Z(k))^{-1}\| \leq n\frac{\bar{\lambda}\bar{c}}{\underline{\lambda}^2\underline{c}^2} = \bar{k}.$$

Thus, it is clear that

$$\|I + \bar{K}(k)\bar{\Gamma}(k)\mathcal{C}(k)\| \leq 1 + \bar{k}\bar{\lambda}\bar{c} \triangleq \bar{b}.$$

Moreover, we have

$$\|\aleph(k)\| \leq \sum_{s=1}^{m} \|\Pi_{s2}(k)\| \|\mathrm{tr}[X(k)\Gamma_s(k)]\| \leq \sum_{s=1}^{m} \kappa_{s2}\kappa_s\bar{\tau}$$

and

$$\begin{aligned}&\|\Xi \odot \mathcal{C}'(k)\mathcal{X}(k)\mathcal{C}'(k)\| \\ &\quad= \left\|\mathrm{diag}_n\{\sigma_{\gamma_i}^2 C_i'(k)X(k)C_i(k)\}\right\| \leq \bar{\sigma}^2\bar{c}^2\bar{\tau}.\end{aligned}$$

Therefore, it is obvious that

$$\begin{aligned}&\|\bar{\Upsilon}(k)\| \\ &\leq \|\aleph(k)\| + \|\mathrm{diag}_n\{R_i(k)\}\| + \|\Xi \odot \mathcal{C}'(k)\mathcal{X}(k)\mathcal{C}'(k)\| \\ &\leq \sum_{s=1}^{m} \kappa_{s2}\kappa_2\bar{\tau} + \bar{\sigma}^2\bar{c}^2\bar{\tau} + \bar{r} \triangleq \bar{h}.\end{aligned} \tag{8.25}$$

By letting $\bar{\zeta} = \max_i \left\{\sum_{s=1}^{n} \delta_{is}\right\}$, we have

$$\begin{aligned}&\|\lambda_{max}(\bar{\Gamma}(k)\mathcal{C}(k)M_{k|k-1}\mathcal{C}'(k)\bar{\Gamma}(k) + \bar{\Upsilon}(k))\Lambda\| \\ &\leq (\bar{\lambda}^2\bar{c}^2\|M_{k|k-1}\| + \bar{h})\bar{\zeta}.\end{aligned}$$

In light of (8.14b), it is straightforward to see that

$$\begin{aligned}\|M_{k|k}\| &\leq (\bar{b}^2 + \bar{\lambda}^2\bar{c}^2\bar{\zeta})\|M_{k|k-1}\| + \bar{k}^2\bar{h} + \bar{h}\bar{\zeta} \\ &\leq \bar{a}^2(\bar{b}^2 + \bar{\lambda}^2\bar{c}^2\bar{\zeta})\|M_{k-1|k-1}\| + \bar{h}(\bar{k}^2 + \bar{\zeta}) \\ &\quad + (\bar{b}^2 + \bar{\lambda}^2\bar{c}^2\bar{\zeta})\left(n\sum_{s=1}^{m} \kappa_{s1}\kappa_s\bar{\tau} + n\bar{q}\right),\end{aligned}$$

where the second inequality comes from substituting (8.23). Since $\bar{a}^2(\bar{b}^2 + \bar{\lambda}^2\bar{c}^2\bar{\zeta}) < 1$, the sequence $\|M_{k|k}\|$ converges eventually. Using the fact that $M_{k|k}$ always is the upper bound of the real estimation error covariance $P_{k|k}$, we conclude that the filtering error dynamics is mean-square stable, which ends the proof.

Remark 8.3 According to (8.14b), it is clear that the gain variations do have a great impact on the covariance $M_{k|k}$. Moreover, it can be seem from condition (8.22) in the above theorem that, since $\lambda_{\max}(\bar{\Gamma}(k)\mathcal{C}(k)M_{k|k-1}\mathcal{C}'(k)\bar{\Gamma}(k) + \bar{\Upsilon}(k))\Lambda$ is a multiplicative term in the proliferation of $M_{k|k}$, the sequence $\{M_{k|k}\}$ will diverge quickly if δ_{ij} is too large. As such, it is observed that a smaller gain variation δ_{ij} is more beneficial for the mean-square boundedness.

8.5 An Illustrative Example

In this section, a numerical example is employed to demonstrate the effectiveness of the proposed distributed resilient filter scheme. A target tracking scenario is used to justify its potential applicability.

Consider a wireless sensor network with $n = 4$ sensor nodes. The network topology is represented by a directed graph $G = (V, E, \mathcal{H})$ as shown in Fig. 8.1 with the set of nodes $V = \{1, 2, 3, 4\}$, the set of edges $E = \{(1, 1), (1, 2), (2, 2), (2, 3), (3, 1), (3, 2), (3, 3), (4, 3)\}$, and the adjacency matrix

$$\mathcal{H} = [a_{ij}]_{4\times4} = \begin{bmatrix} 1 & 0.3 & 0 & 0.1 \\ 0 & 1 & 0.2 & 0 \\ 0.5 & 1 & 1 & 0 \\ 0 & 0 & 0.8 & 1 \end{bmatrix}.$$

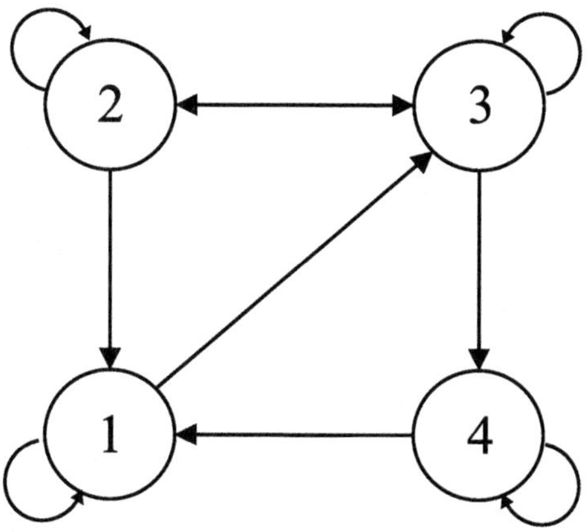

Fig. 8.1 The topology of the WSN

The discretized target system (8.1) with stochastic nonlinearities is described by

$$x(k+1) = \begin{bmatrix} 0.89 & 0.1+0.1\cos(0.12k) \\ 0 & 0.88 \end{bmatrix} x(k) + w(k) + f(k, x(k), \xi(k)).$$

The initial values of the state $x(0)$ and the process noise $w(k)$ follow the zero-mean Gaussian distribution with the respective covariances $\Sigma_0 = \text{diag}\{2, 2\}$ and $Q(k) = \text{diag}\{0.1, 0.15\}$. The parameters of the measurement models of the sensors (8.2) are described as follows:

$$\begin{aligned}
C_1(k) &= [0.92 + 0.05\cos(0.12k) \;\; 0.82], \\
C_2(k) &= [0.25 \;\; 0.1 + 0.05\sin(0.1k)], \\
C_3(k) &= [0.84 + 0.05\cos(0.1k) \;\; 0.75 + 0.05\sin(0.1k)], \\
C_4(k) &= [0.75 \;\; 0.435].
\end{aligned}$$

Suppose that the stochastic variables $v_i(k)$ are independent zero-mean Gaussian white noise sequences with the covariances $R_i(k) = 0.25,\ i = 1, 2, 3, 4$. The stochastic sensor gain degradation of individual sensors has the following probability density function:

$$p_k^i(s) = \begin{cases} 0.05, & s = 0 \\ 0.10, & s = 0.5 \\ 0.85, & s = 1 \end{cases}$$

for $i = 1, 2, \ldots, 4$. As such, the expectation and variance can be easily calculated as $\bar{\gamma}_i(k) = 0.9$ and $\sigma_{\gamma_i}^2(k) = 0.065$, respectively. The stochastic nonlinearities $f(k, x(k), \xi(k))$ and $g_i(k, x(k), \zeta_i(k))$ are selected as follows:

$$\begin{aligned}
f(k, x(k), \xi(k)) &= \begin{bmatrix} 0.1 \\ 0.2 \end{bmatrix} (0.3\text{sign}(x^{(1)}(k))x^{(1)}(k)\xi^{(1)}(k) \\
&\quad + 0.4\text{sign}(x^{(2)}(k))x^{(2)}(k)\xi^{(2)}(k)) \\
g_i(k, x(k), \zeta_i(k)) &= 0.3(0.3\text{sign}(x^{(1)}(k))x^{(1)}(k)\zeta_i^{(1)}(k) \\
&\quad + 0.4\text{sign}(x^{(2)}(k))x^{(2)}(k)\zeta_i^{(2)}(k)),
\end{aligned}$$

where $x^{(j)}(k)$, $\xi^{(j)}(k)$, and $\zeta_i^{(j)}(k)$ $(j = 1, 2)$ denote the jth elements of the system state $x(k)$, and the stochastic variables $\xi(k)$ and $\zeta_i(k)$, respectively. Obviously, the expectations and the covariances of the above stochastic nonlinearities meet the form in (8.3) and (8.4) with the integer $m = 1$, parameter matrices $\Pi_{s1} = [0.1\ 0.2]'[0.1\ 0.2]$, $\Pi_{s2}^{(i)} = 0.09$, and $\Gamma_s(k) = \text{diag}\{0.09,\ 0.16\}$. The initial parameters of the filters are chosen as $\hat{x}_i(0|0) = 0$ and $M_{0|0} = \mathbf{1}_4 \otimes \Sigma_0$. Additionally, assume that $\delta_{ij} = 0.1$, for $i, j = 1, \ldots, n$. We can compute the filter gain parameters

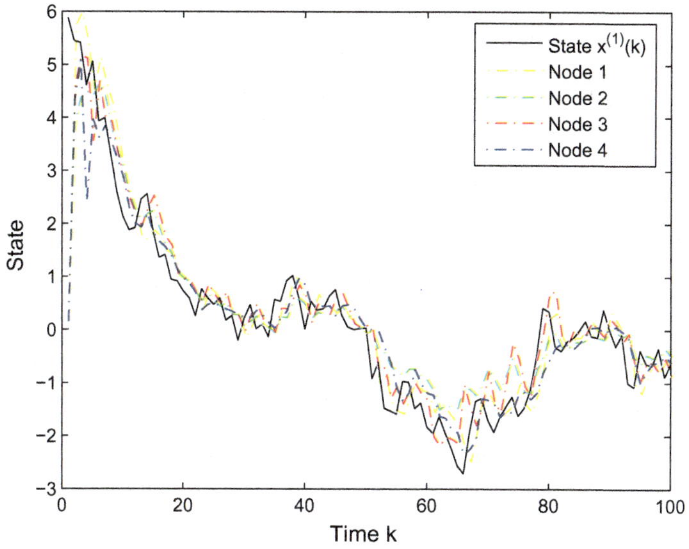

Fig. 8.2 The true state $x^{(1)}(k)$ and its estimates $\hat{x}_i^{(1)}(k|k)$

according to (8.14a), (8.14b), and (8.18), and then exploit the algorithm given by (8.5a)–(8.5b) to estimate the state vector in a distributed manner.

The simulation results are presented in Figs. 8.2, 8.3, 8.4, 8.5 and 8.6. Among them, Figs. 8.2 and 8.3 depict the trajectories of the true states $x^{(j)}(k)$ and the corresponding estimates $\hat{x}_i^{(j)}(k|k)$. Figures 8.4 and 8.5 give the values of the estimation errors of the sensor nodes, respectively. To quantify the estimation accuracy, the mean-square estimation error is defined as follows:

$$\mathbf{MSE}(k) = \frac{1}{T}\sum_{t=1}^{T}\sum_{i=1}^{n} e_i'(k|k)e_i(k|k).$$

Figure 8.6 presents the trace of the matrix $M_{k|k}$ calculated from Theorem 8.3.1 and the mean-square error (MSE) obtained from $T = 1,000$ independent experiments. The result confirms that the solutions of the difference Eq. (8.14a)–(8.14b) are actually the upper bounds of the error variance. Moreover, we compare the MSE of our resilient distributed filter with that of the filter proposed in [25]. Form the simulation results in Fig. 8.7, it can be seen that our resilient distributed filter performs better, which is not surprising as we have made specific efforts to account for the stochastic gain variations, the nonlinearities, and the sensor gain degradation.

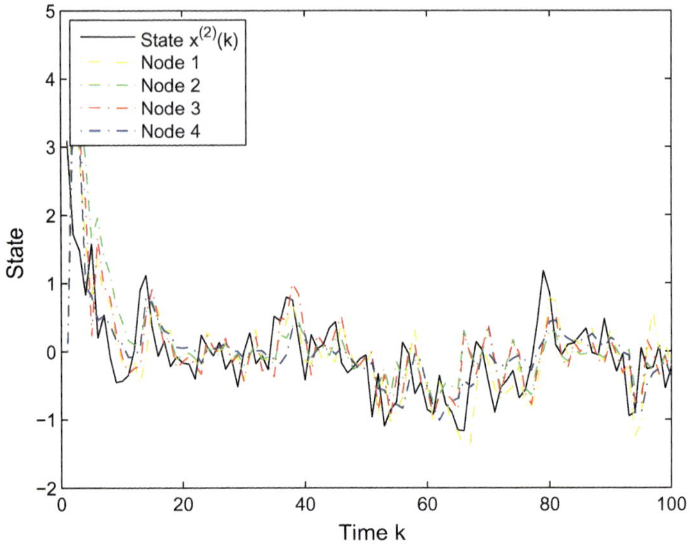

Fig. 8.3 The true state $x^{(2)}(k)$ and its estimates $\hat{x}_i^{(2)}(k|k)$

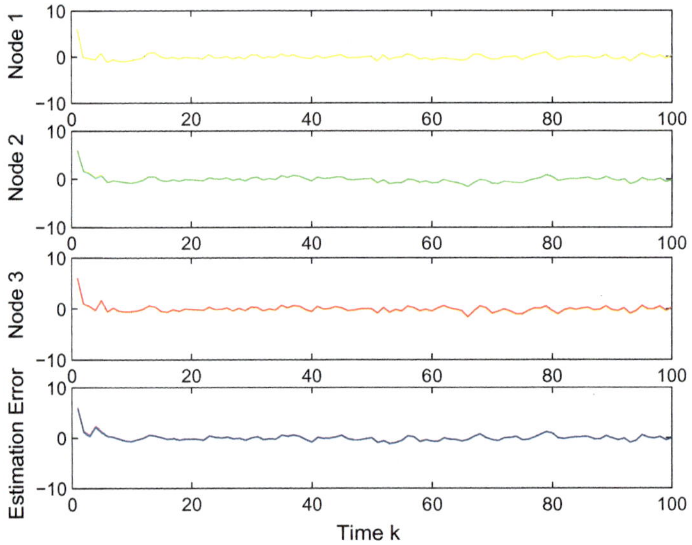

Fig. 8.4 The estimation error of $x^{(1)}(k)$

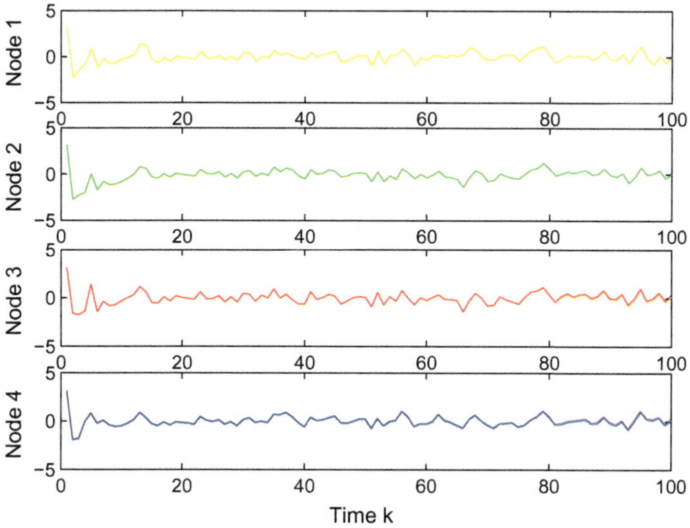

Fig. 8.5 The estimation error of $x^{(2)}(k)$

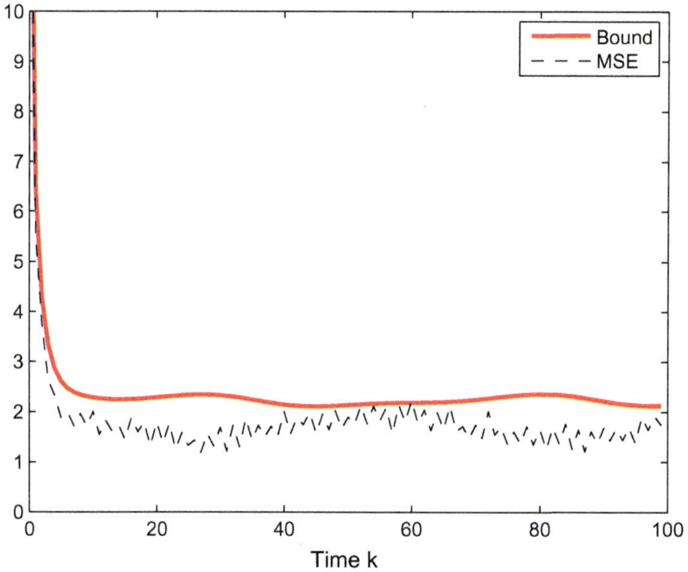

Fig. 8.6 The MSE and its upper bound $\mathrm{tr}\{M_{k|k}\}$

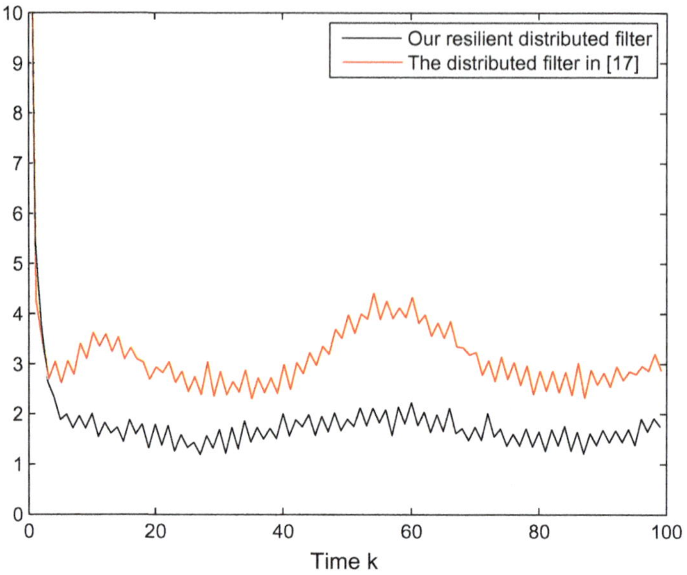

Fig. 8.7 MSE Comparison for the proposed resilient filter and the filter in [25]

8.6 Conclusions

In this chapter, we have investigated the distributed filtering problem for discrete time-varying systems subject to complicated stochastic phenomena including stochastic nonlinearities, sensor degradation, and gain variations. In the presence of these stochastic phenomena, it is impossible to obtain the exact error covariance in an explicit form, let alone the design of the filter gains. To tackle this problem, a suboptimal distributed resilient filter design scheme has been established. Specifically, we have derived a matrix difference equation whose solution is the upper bound of the actual error covariance. Filter gains have been designed through minimizing such an upper bound at each step iteratively. After that, a sufficient condition has been established to guarantee the mean-square stability of the distributed resilient filter. Finally, the effectiveness of the proposed filtering algorithm has been illustrated by a numerical example.

References

1. M.V. Basin, J.J. Maldonado, H.R. Karimi, Mean-square filtering for polynomial system states confused with poisson noises over polynomial observations. Model. Identif. Control **32**(2), 47–55 (2011)
2. R. Caballero-Águila, A. Hermoso-Carazo, J. Linares-Pérez, Optimal state estimation for networked systems with random parameter matrices, correlated noises and delayed measurements. Int. J. Gen. Syst. **44**(2), 142–154 (2015)

3. H.R. Karimi, A linear matrix inequality approach to robust fault detection filter design of linear systems with mixed time-varying delays and nonlinear perturbations. J. Frankl Inst. **347**(6), 957–973 (2010)
4. S. Sun, L. Xie, W. Xiao, Optimal full-order and reduced-order estimators for discrete-time systems with multiple packet dropouts. IEEE Trans. Signal Process. **56**(8), 4031–4038 (2008). Aug
5. H. Wang, G. Yang, Robust mixed l_1/H_∞ filtering for affine fuzzy systems with measurement errors. IEEE Trans. Cybern. **44**(7), 1100–1110 (2014)
6. X. Yao, L. Wu, W. Zheng, Fault detection filter design for markovian jump singular systems with intermittent measurements. IEEE Trans. Signal Process. **59**(7), 3099–3109 (2011)
7. F. Yang, Y. Li, Set-membership filtering for systems with sensor saturation. Automatica **45**(8), 1896–1902 (2009)
8. M.S. Mahmoud, Distributed estimation based on information-based covariance intersection algorithms. Int. J. Adapt. Control Signal Process. **30**, 750–778 (2016)
9. M.S. Mahmoud, M.F. Emzir, State estimation with asynchronous multi-rate multi-smart sensors. Inf. Sci. **196**, 15–27 (2012)
10. Q. Li, B. Shen, Y. Liu, F.E. Alsaadi, Event-triggered H_∞ state estimation for discrete-time stochastic genetic regulatory networks with Markovian jumping parameters and time-varying delays. Neurocomputing **174**, 912–920 (2016)
11. R. Olfati-Saber, R.M. Murray, Consensus problems in networks of agents with switching topology and time-delays. IEEE Trans. Autom. Control **49**(9), 1520–1533 (2004). Sep
12. E. Yaz, Y. Yaz, State estimation of uncertain nonlinear stochastic systems with general criteria. Appl. Math. Lett. **14**(5), 605–610 (2001)
13. J. Hu, Z. Wang, H. Gao, Recursive filtering with random parameter matrices, multiple fading measurements and correlated noises. Automatica **49**(2013), 3440–3448 (2013)
14. J. Song, Y. Niu, Resilient finite-time stabilization of fuzzy stochastic systems with randomly occurring uncertainties and randomly occurring gain fluctuations. Neurocomputing **171**, 444–451 (2016)
15. A. Manor, A. Osovizky, E. Dolev, E. Marcus, D. Ginzburg, V. Pushkarsky, Y. Kadmon, Y. Cohen, Compensation of scintillation sensor gain variation during temperature transient conditions using signal processing techniques. *In IEEE Nuclear Science Symposium Conference, Orlando*, FL, USA, October, pp. 2399–2403 (2009)
16. X. Wang, E. Yaz, Stochastically resilient extended Kalman filtering for discrete-time nonlinear systems with sensor failures. Int. J. Syst. Sci. **45**(7), 1393–1401 (2014)
17. F. Hounkpevi, E. Yaz, Robust minimum variance linear state estimators for multiple sensors with different failure rates. Automatica **43**(7), 1274–1280 (2007)
18. J. Song, Y. Niu, S. Wang, Robust finite-time dissipative control subject to randomly occurring uncertainties and stochastic fading measurements. J. Frankl. Inst. (2016). https://doi.org/10.1016/j.jfranklin.2016.07.020
19. G. Yang, J. Wang, Robust nonfragile kalman filtering for uncertain linear systems with estimator gain uncertainty. IEEE Trans. Automatic Control **46**(2), 343–348 (2001)
20. M.S. Mahmoud, Resilient linear Filtering of uncertain systems. Automatica **40**(10), 1797–1802 (2004)
21. G. Yang, W. Che, Non-fragile H_∞ filter design for linear continuous-time systems. Automatica **44**(11), 2849–2856 (2008)
22. G. Yang, X. Guo, Insensitive H_∞ filter design for continuous-time systems with respect to filter coefficient variations. Automatica **46**(11), 1860–1869 (2010)
23. N. Hou, H. Dong, Z. Wang, W. Ren, E. Fuad, Alsaadi non-fragile state estimation for discrete Markovian jumping neural networks. Neurocomputing **179**, 238–245 (2016)
24. R. Horn, C. Johnson, *Topics in matrix analysis* (Cambridge University Press, New York, 1991)
25. Y. Liu, Z. Wang, X. He, D.H. Zhou, Minimum-variance recursive filtering over sensor networks with stochastic sensor gain degradation: algorithms and performance analysis. IEEE Trans. Control Netw. Syst. **3**(3), 265–274 (2016)

Chapter 9
Consensus-Based Recursive Distributed Filtering

A sensor network is composed of a large number of spatially distributed sensor nodes working together to monitor targets cooperatively. In recent years, sensor networks have stirred a great deal of research attention because of their great potential for many applications including seismic sensing, environmental monitoring, and military target tracking, see [1] for a survey. Compared with a single sensor, in distributed filtering schemes, the local filters estimate the system state based on the information not only from itself but also from its adjacent sensors according to the topology of networks. A challenging problem with sensor networks is how to design an appropriate information fusion scheme in order to characterize the information exchange among the sensor nodes.

In recent years, many results have been reported with respect to distributed filtering problems for complex/sensor networks, see, e.g., [2, 3]. Specifically, the distributed state estimation problem has been considered in [2] for systems with randomly occurring phenomena and, by utilizing linear matrix inequalities, the estimator gains have been appropriately designed to guarantee that the disturbance attenuation meets the H_∞ performance index. In [3], the distributed H_∞ state estimation problem has been addressed for discrete-time Markovian jump nonlinear time-delay systems with incomplete statistics of mode transitions. Furthermore, based on the average consensus schemes, the distributed estimation problems have been investigated in many papers such as [4, 5]. In [6], the authors have generated the local estimates by a one-step consensus, which is actually a convex combination between the local and the adjacent intermediate estimates. Another work in [5] has developed an average consensus strategy to obtain the weighted least squares approximate solution of static systems over the networks. Besides, the optimal decentralized Kalman-consensus filter for linear discrete-time systems has been proposed in [4].

It is well known that nonlinearities inevitably exist in almost all practical systems, and the existence of the nonlinearities, if not properly handled, might lead to

Q. Liu et al., *Stochastic Control and Filtering over Constrained Communication Networks*, Studies in Systems, Decision and Control 178,
https://doi.org/10.1007/978-3-030-00157-5_9

the degradation of system performance or even cause instability. As a result, analysis and synthesis problems for nonlinear systems have received particular research attention over the past few decades. Note that the nonlinearities studied in the majority of available literature have been assumed to occur in a deterministic way. Nonetheless, sometimes, the nonlinearities might be originated due to randomly occurring phenomena such as sudden environment changes. In fact, the widely investigated state-multiplicative noises and stochastic vectors whose power depends on the sign of the states are both the special cases of stochastic nonlinearities. Up to now, only a limited number of results have been available in the literature on general stochastic nonlinearities, see, e.g., [7–9], but the distributed filtering problem with general stochastic nonlinearities in both the target plant and the sensor model has not been adequately researched yet, and this gives the primary motivation of the present investigation.

Summarizing the aforementioned discussion, in this chapter, we aim to address the distributed filtering problem for linear discrete time-varying systems with stochastic nonlinearities characterized by statistical information, that is, the first and second-order moments. In order to accelerate the information fusion among sensor nodes, an extra consensus step is introduced to average the estimates in the networks. The main contributions of this chapter can be highlighted as follows. (1) *The general stochastic nonlinearities are covered in both the process and the measurements.* (2) *Consensus-based Kalman-like filters are introduced to restore the state of the process in a distributed manner.* (3) *Suboptimal filter gains are designed to minimize the upper bound of the filtering error covariance by recursive calculations facilitating online applications.*

9.1 Problem Formulation and Preliminaries

9.1.1 *Target Plant and Sensor Network*

Consider the following class of discrete time-varying systems with stochastic nonlinearity:

$$x(k+1) = A(k)x(k) + f(k, x(k), \xi(k)) + B(k)w(k), \tag{9.1}$$

where $x(k) \in \mathbb{R}^{n_x}$ is the state vector that cannot be measured directly, and $w(k) \in \mathbb{R}^{n_w}$ is a sequence of Gaussian random variables with zero-mean value and covariance matrix $Q(k)$. $A(k)$ and $B(k)$ are known time-varying matrices of appropriate dimensions. The initial condition $x(0)$ is assumed to obey a Gaussian distribution with mean μ_0 and covariance matrix Σ_0.

In this chapter, a sensor network consisting of n sensor nodes is exploited to measure the output of the target plant. We denote the topology of the network by a directed graph $G = (V, E, A)$ of order n with the set of nodes $V = \{1, 2, \ldots, n\}$, the set of edges $E \subseteq V \times V$, and the weighted adjacency matrix $A = [a_{ij}]_{n\times n}$. The weighted adjacency matrix of the graph is a matrix with nonnegative elements a_{ij} satisfying the property $a_{ij} > 0 \Longleftrightarrow (i, j) \in E$, which means that the ith node can receive the information from the jth node. The Laplacian matrix $H = D - A$, where diagonal matrix $D = \text{diag}\{d_1, d_2, \ldots, d_n\}$ is the degree of matrix A with elements $d_i = \sum_{j=1}^{n} a_{ij}$. All the neighbors of node i are denoted by the set as $N_i \triangleq \{j \in V|(i, j) \in E\}$. The graph G is assumed to have a spanning tree, such that there is a directed tree formed by graph edges that connect all the nodes of the graph.

For the ith sensor node, the model is described by

$$y_i(k) = C_i(k)x(k) + g_i(k, x(k), \zeta_i(k)) + D_i(k)v_i(k), \tag{9.2}$$

for $i = 1, 2, \ldots, n$, where $y_i(k) \in \mathbb{R}^{n_y}$ stands for the measurement information from sensor i, and the measure noise $v_i(k) \in \mathbb{R}^{n_v}$ is assumed to be Gaussian with zero-mean value and covariance matrices $R_i(k)$. $C_i(k)$ and $D_i(k)$ are known time-varying matrices of appropriate dimensions.

The stochastic variables $\xi(k)$, $\zeta_i(k)$, $w(k)$ and $v_i(k)$ are mutually independent. As in [9], we assume that the functions $f(k, x(k), \xi(k)) \in \mathbb{R}^{n_x}$ and $g_i(k, x(k), \zeta_i(k)) \in \mathbb{R}^{n_y}$ represent the stochastic nonlinearities with $f(k, 0, \xi(k)) = 0$ and $g_i(k, 0, \zeta_i(k)) = 0$, and have the following first-order moment for $x(k)$:

$$\begin{aligned} &\mathbb{E}\{f(k, x(k), \xi(k))|x(k)\} = 0 \\ &\mathbb{E}\{g_i(k, x(k), \zeta_i(k))|x(k)\} = 0 \end{aligned} \tag{9.3}$$

as well as the following covariance:

$$\begin{aligned} &\mathbb{E}\left\{g_i(k, x(k), \zeta_i(k))g_j^T(k, x(k), \zeta_j(k))|x(k)\right\} = 0, \quad i \neq j \\ &\mathbb{E}\left\{\begin{bmatrix} f(k, x(k), \xi(k)) \\ g_i(k, x(k), \zeta_i(k)) \end{bmatrix} \begin{bmatrix} f(s, x(s), \xi(s)) \\ g_i(s, x(s), \zeta_i(s)) \end{bmatrix}^T \middle| x(k)\right\} = 0, \\ &\qquad k \neq s \\ &\mathbb{E}\left\{\begin{bmatrix} f(k, x(k), \xi(k)) \\ g_i(k, x(k), \zeta_i(k)) \end{bmatrix} \begin{bmatrix} f(k, x(k), \xi(k)) \\ g_i(k, x(k), \zeta_i(k)) \end{bmatrix}^T \middle| x(k)\right\} \\ &\qquad = \sum_{s=1}^{m} \Pi_s(k)x^T(k)\Gamma_s(k)x(k), \end{aligned} \tag{9.4}$$

where m is a given positive integer, and $\Pi_s(k) \triangleq \text{diag}\{\Pi_{s1}(k), \Pi_{s2}(k)\}$. $\Pi_{s1}(k)$, $\Pi_{s2}(k)$, and $\Gamma_s(k)$ are known matrices with compatible dimensions for $s = 1, 2, \ldots, m$.

Remark 9.1 The stochastic nonlinearities $f(k, x(k), \xi(k))$ and $g_i(k, x(k), \zeta_i(k))$ introduced in this chapter are quite general, which include some commonly investigated stochastic nonlinear functions as special cases. Examples include state-multiplicative noises $D(k)x(k)w(k)$ and random vectors whose power relies on the sign of the state $\text{sgn}(x(k))Ax(k)w(k)$ where the sign denotes the signum function.

9.1.2 Consensus-Based Filtering Algorithm

Next, we aim to design recursive distributed estimator structures and an information fusion strategy for sensor networks.

In every sensor node, the distributed estimators that only have access to the information from its neighboring nodes according to the communication topology are constructed to extract the state vector of the system. The distributed filtering problem is solved in the following two steps.

Step 1. First, the following Kalman filter structure is adopted on sensor node i:

$$\varphi_i(k) = A(k)\hat{x}_i(k) + K_i(k)(y_i(k) - C_i(k)\hat{x}_i(k)), \tag{9.5}$$

where $\varphi_i(k) \in \mathbb{R}^{n_x}$ is the intermediate estimation variable, $\hat{x}_i(k) \in \mathbb{R}^{n_x}$ represents the estimate of the state of the process, and $K_i(k)$ denotes the filtering gain to be designed.

Step 2. Second, define an auxiliary variable $h_i(k, l)$ with the initial value $h_i(k, 0) = \varphi_i(k)$. An iterative form of the consensus algorithm is given as

$$h_i(k, l+1) = h_i(k, l) + \epsilon \sum_{j \in N_i} a_{ij}\big(h_j(k, l) - h_i(k, l)\big) \tag{9.6}$$

for $l = 1, 2, \ldots, L$, where L is the prespecified consensus step and ϵ is a positive scalar.

Subsequently, by letting $h(k, l) = [h_1^T(k, l), h_2^T(k, l), \ldots, h_n^T(k, l)]^T$ and Perron matrix $P = I - \epsilon H$, the discrete-time collective dynamics of (9.6) can be written as follows:

$$h(k, l+1) = Ph(k, l)$$

by which one has $h(k, L) = P^L h(k, 0)$. Partitioning the matrix P^L as $\{p_{ij}\}_{n\times n}$, we obtain

$$h_i(k, L) = \sum_{j=1}^{n} p_{ij} h_j(k, 0)$$

for $i = 1, 2, \ldots, n$. Letting the estimation of sensor i at the $(k+1)$th sampling step be $\hat{x}_i(k+1) = h_i(k, L)$, we arrive at

$$\hat{x}_i(k+1) = \sum_{j=1}^{n} p_{ij}[A(k)\hat{x}_j(k) + K_j(k)(y_j(k) - C_j(k)\hat{x}_j(k))] \tag{9.7}$$

Lemma 9.1.1 ([10]) *If a nonnegative matrix $P = [\tilde{p}_{ij}]_{n\times n}$ has the same positive constant row sum $\alpha > 0$, then α is an eigenvalue of P with an associated eigenvector $\mathbf{1}$ and spectral radius $\rho(P) = \alpha$. Besides, the eigenvalue α of P has algebraic multiplicity equal to one if and only if the graph associated with P has a spanning tree. Furthermore, if the graph associated with P has a spanning tree and $\tilde{p}_{ii} > 0$, then α is the unique eigenvalue of maximum modulus.*

A nonnegative matrix is said to be row stochastic if all of its row sums are 1. It is assumed in this chapter that the graph G has a spanning tree and the parameter $\epsilon \in (0, 1/\Delta]$, where Δ is the maximum degree of the weighted adjacency matrix A, i.e., $\Delta = \max_i(\sum_{j\neq i} a_{ij})$. Therefore, it follows from Lemma 1 that the nonnegative matrix P is a stochastic matrix with algebraic multiplicity equal to one for its eigenvalue 1. Additionally, it is not difficult to verify that the nonnegative matrix P^L is also a stochastic matrix, i.e., $\{p_{ij}\}_{j=1}^{n}$ sums up to one.

Remark 9.2 The value of L determines the speed of information fusion. When it comes to the situation that $L = 0$, that is, $P^L = I$, the consensus-based distributed filters reduces to the traditional Kalman filters. On the other hand, according to the work in [10], $\lim_{l\to\infty} P^l \to \mathbf{1}\nu^{\mathrm{T}}$, where ν satisfies $P^T\nu = \nu$, $\mathbf{1}^{\mathrm{T}}\nu = \mathbf{1}$, and each element of ν is nonnegative. As such, if the consensus step L tends to infinity, the estimate $\hat{x}_i(k)$ is equal to each other.

By defining the estimation error $\tilde{x}_i(k) = x_i(k) - \hat{x}_i(k)$ and introducing the covariance matrix of the estimation error of the ith sensor as $\Omega_i(k) = \mathbb{E}\{\tilde{x}_i(k)\tilde{x}_i^T(k)\}$, the following error dynamics for the sensor networks can be obtained from (9.1) and (9.7):

$$\begin{aligned} &\tilde{x}_i(k+1) \\ &= \sum_{j=1}^{n} p_{ij}[(A(k) - K_j(k)C_j(k))\tilde{x}_j(k) + f(k, x(k), \xi(k)) \\ &\quad + B(k)w(k) - K_j(k)g_j(k, x(k), \zeta_j(k)) \\ &\quad - K_j(k)D_j(k)v_j(k)], \end{aligned} \tag{9.8}$$

with initial estimate $\hat{x}_i(0) = \mu_0$ and estimation covariance $\Omega_i(0) = \Sigma_0$. In the following analysis, we intend to design the filter gain matrices in order to locally minimize an upper bound of estimation error covariance for the distributed filtering system.

9.2 Main Results

Before proceeding further, we need to introduce the following lemma that will be utilized in the subsequent analysis.

Lemma 9.2.1 *Consider the discrete time-varying process (9.1) with stochastic nonlinearities. The system state covariance* $X(k) = \mathbb{E}\{x(k)x^T(k)\}$ *satisfies the following recursive equation:*

$$\begin{aligned} X(k+1) = & A(k)X(k)A^T(k) + B(k)Q(k)B^T(k) \\ & + \sum_{s=1}^{m} \Pi_{s1}(k)\mathrm{tr}\{X(k)\Gamma_s(k)\} \end{aligned} \tag{9.9}$$

with the initial value $X(0) = \Sigma_0 + \mu_0\mu_0^T$

Proof By substituting $x(k+1)$ of (9.1) into $X(k+1)$ and taking the expectation with respect to $w(k)$ and $\xi(k)$, we obtain

$$\begin{aligned} & X(k+1) \\ & = \mathbb{E}\{(A(k)x(k) + f(k, x(k), \xi(k)) + B(k)w(k)) \\ & \quad (A(k)x(k) + f(k, x(k), \xi(k)) + B(k)w(k))^T\} \\ & = A(k)\mathbb{E}\{x(k)x^T(k)\}A^T(k) + B(k)\mathbb{E}\{w(k)w^T(k)\} \\ & \quad \times B^T(k) + \mathbb{E}\{f(k, x(k), \xi(k))f^T(k, x(k), \xi(k))\} \\ & = A(k)X(k)A^T(k) + B(k)Q(k)B^T(k) \\ & \quad + \sum_{s=1}^{m} \Pi_{s1}(k)\mathrm{tr}\{X(k)\Gamma_s(k)\} \end{aligned}$$

Since the initial value $x(0)$ is assumed to be Gaussian with mean μ_0 and covariance matrix Σ_0, that is, $X(0) = \mathbb{E}\{(x(0) - \mu_0)(x(0) - \mu_0)^T\} = \Sigma_0$, it can be concluded that $X(0) = \Sigma_0 + \mu_0\mu_0^T$, which ends the proof.

Lemma 9.2.2 ([6]) *Given a positive integer* n*, a set of vectors* $\{x_i\}_{i=1}^n$*, a set of matrices* $\{A_i\}_{i=1}^n$*, and a set of nonnegative scalars* $\{p_i\}_{i=1}^n$ *summing up to one. The following inequality holds:*

$$(\sum_{i=1}^{n} p_i A_i x_i)(\sum_{i=1}^{n} p_i A_i x_i)^T \leq \sum_{i=1}^{n} p_i A_i x_i x_i^T A_i^T \tag{9.10}$$

Theorem 9.2.3 *Consider the discrete time-varying stochastic process (9.1) with consensus-based distributed filters (9.5) and (9.6). If a set of positive definite matrices* $\{\Xi_i(k)\}_{k=1}^{N}$ *solves the following Riccati-like equation:*

$$
\begin{aligned}
&\Xi_i(k+1) \\
&= \sum_{j=1}^{n} p_{ij}(A(k) - K_j(k)C_j(k))\Xi_j(k)(A(k) - K_j(k) \\
&\times C_j(k))^T + \sum_{s=1}^{m} \Pi_{s1}(k)\mathrm{tr}\{X(k)\Gamma_s(k)\} + B(k)Q(k) \\
&\times B^T(k) + \sum_{j=1}^{n} p_{ij}K_j(k)[\sum_{s=1}^{m} \Pi_{s2}(k)\mathrm{tr}\{X(k)\Gamma_s(k)\} \qquad (9.11)\\
&+ D_j(k)R_j(k)D_j^T(k)]K_j^T(k), \qquad (9.12)
\end{aligned}
$$

with $\Xi_i(0) = \Omega_i(0) \geq 0$ $(i = 1, 2, \ldots, n)$, *then* $\Xi_i(k)$ *is the upper bound of the estimation error covariance matrix* $\Omega_i(k)$, *i.e.,* $\Xi_i(k) \geq \Omega_i(k)$, *for* $k = 1, 2, \ldots, N$.

Proof The estimation error covariance could be computed as follows:

$$
\begin{aligned}
&\Omega_i(k+1) \\
&= \mathbb{E}\Big\{(\sum_{j=1}^{n} p_{ij}(A(k) - K_j(k)C_j(k))\tilde{x}_j(k))(\sum_{j=1}^{n} p_{ij}(A(k) \\
&- K_j(k)C_j(k))\tilde{x}_j(k))^T + B(k)w(k)w^T(k)B^T(k) \\
&+ f(k, x(k), \xi(k))f^T(k, x(k), \xi(k)) + (\sum_{j=1}^{n} p_{ij}K_j(k) \\
&\times g_j(k, x(k), \zeta_j(k))(\sum_{j=1}^{n} p_{ij}K_j(k)g_j(k, x(k), \zeta_j(k))^T \\
&+ (\sum_{j=1}^{n} p_{ij}K_j(k)D_j(k)v_j(k))(\sum_{j=1}^{n} p_{ij}K_j(k)D_j(k)v_j(k))^T\Big\}
\end{aligned}
$$

Using the inequality (9.10) in Lemma 9.2.2 and the properties of the stochastic nonlinearities in (9.4), we obtain

$$
\begin{aligned}
&\Omega_i(k+1) \\
&\leq \sum_{j=1}^{n} p_{ij}(A(k) - K_j(k)C_j(k))\Omega_j(k)(A(k) - K_j(k)C_j(k))^T \\
&+ \sum_{s=1}^{m} \Pi_{s1}(k)\mathrm{tr}\{X(k)\Gamma_s(k)\} + B(k)Q(k)B^T(k) \\
&+ \sum_{j=1}^{n} p_{ij}K_j(k)[\sum_{s=1}^{m} \Pi_{s2}(k)\mathrm{tr}\{X(k)\Gamma_s(k)\} \\
&+ D_j(k)R_j(k)D_j^T(k)]K_j^T(k).
\end{aligned}
$$

Assuming that $\Omega_i(k) \leq \Xi_i(k)$, for all $i = 1, 2, \ldots, n$, it follows that

$$
\begin{aligned}
\Omega_i(k+1) - \Xi_i(k+1) \leq \sum_{j=1}^{n} p_{ij}(A(k) - K_j(k)C_j(k)) \\
\times [\Omega_j(k) - \Xi_j(k)](A(k) - K_j(k)C_j(k))^T \leq 0
\end{aligned}
\tag{9.13}
$$

and $\Omega_i(k+1) \leq \Xi_i(k+1)$. Note that $\Xi_i(0) = \Omega_i(0)$, by inductive method, one can conclude from (9.12) and (9.13) that $\Omega_i(k) \leq \Xi_i(k)$, for $i = 1, 2, \ldots, n$ and $k = 1, 2, \ldots, N$. The proof is complete.

Up to now, the upper bounds of the estimation error covariance are obtained by recursively computing the Riccati-like equation (9.12). In the sequel, we address the problem of designing suboptimal filter gains to minimize the upper bounds $\Xi_i(k)$.

Theorem 9.2.4 *Consider the discrete time-varying stochastic process (9.1) with consensus-based distributed filters (9.5) and (9.6). Given a set of system state covariance matrices $\{X(k)\}_{k=1}^N$ and the upper bound of estimation error covariance matrices $\{\Xi_i(k)\}_{k=1}^N$ computed by (9.9) and (9.12), respectively. The suboptimal filtering gains are designed as follows:*

$$
K_i^*(k) = A(k)\Xi_i^*(k)C_i^T(k)\Upsilon_i^{*-1}(k), \quad \text{for } i = 1, 2, \ldots, n, \tag{9.14}
$$

where $\Upsilon_i^*(k) = \sum_{s=1}^m \Pi_{s2}(k)\mathrm{tr}\{X(k)\Gamma_s(k)\} + D_i(k)R_i(k)D_i^T(k) + C_i(k)\Xi_i^*(k)C_i^T(k)$ *and the locally minimized upper bound of the estimation error covariance is calculated using*

$$
\begin{aligned}
&\Xi_i^*(k+1) \\
&= \sum_{j=1}^{n} p_{ij}[A(k)\Xi_j^*(k)A^T(k) - A(k)\Xi_j^*(k)C_j^T(k)\Upsilon_j^{*-1}(k) \\
&\times C_j(k)\Xi_j^*(k)A^T(k)] + \sum_{s=1}^{m} \Pi_{s1}(k)\mathrm{tr}\{X(k)\Gamma_s(k)\} \\
&+ B(k)Q(k)B^T(k)
\end{aligned}
\tag{9.15}
$$

with $\Xi_i^*(0) = \Omega_i(0) \geq 0\ (i = 1, 2, \ldots, n)$.

Proof According to (9.12), we have

$$
\begin{aligned}
&\Xi_i(k+1) \\
&= \sum_{j=1}^{n} p_{ij}A(k)\Xi_j(k)A^T(k) - \sum_{j=1}^{n} p_{ij}K_j(k)C_j(k)\Xi_j(k) \\
&\times A^T(k) - \sum_{j=1}^{n} p_{ij}A(k)\Xi_j(k)C_j^T(k)K_j^T(k) \\
&+ \sum_{s=1}^{m} \Pi_{s1}(k)\mathrm{tr}\{X(k)\Gamma_s(k)\} + B(k)Q(k)B^T(k) \\
&+ \sum_{j=1}^{n} p_{ij}K_j(k)[\sum_{s=1}^{m} \Pi_{s2}(k)\mathrm{tr}\{X(k)\Gamma_s(k)\} \\
&+ D_j(k)R_j(k)D_j^T(k) + C_j(k)\Xi_j(k)C_j^T(k)]K_j^T(k).
\end{aligned}
$$

Completing the squares gives

$$
\begin{aligned}
&\Xi_i(k+1) \\
&= \sum_{j=1}^{n} p_{ij}[A(k)\Xi_j(k)A^T(k) - A(k)\Xi_j(k)C_j^T(k)\Upsilon_j^{-1}(k) \\
&\times C_j(k)\Xi_j(k)A^T(k)] + \sum_{j=1}^{n} p_{ij}(K_j(k) - A(k)\Xi_j(k) \\
&\times C_j^T(k)\Upsilon_j^{-1}(k))\Upsilon_j(k)(K_j(k) - A(k)\Xi_j(k)C_j^T(k) \\
&\times \Upsilon_j^{-1}(k))^T + \sum_{s=1}^{m} \Pi_{s1}(k)\mathrm{tr}\{X(k)\Gamma_s(k)\} \\
&+ B(k)Q(k)B^T(k).
\end{aligned}
$$

Clearly, the suboptimal strategy for filter gains $K_i(k)$ is $K_i^*(k) = A(k)\Xi_i^*(k)C_i^T(k)$ $\Upsilon_i^{*-1}(k)$, which ends the proof.

9.3 An Illustrative Example

In this section, a numerical example is provided to verify the effectiveness of the designed consensus-based filters.

Consider a sensor network consisting of $n = 4$ sensor nodes, which is represented by a directed graph $G = (V, E, A)$ with the set of nodes $V = \{1, 2, 3, 4\}$, the set of edges $E = \{(1, 2), (2, 3), (3, 1), (3, 4), (4, 3)\}$, and the adjacency matrix associated with the edges of the graph

$$A = \begin{bmatrix} 0 & 0.6 & 0 & 0 \\ 0 & 0 & 0.7 & 0 \\ 0.5 & 0 & 0 & 1 \\ 0 & 0 & 1.8 & 0. \end{bmatrix}$$

The discrete time-varying system (9.1) under consideration is described by

$$\begin{aligned} x(k+1) = &\begin{bmatrix} 1 & 0.1 + 0.1\cos(0.12k) \\ 0 & 0.88 \end{bmatrix} x(k) + \begin{bmatrix} 0.16 \\ 0.18 \end{bmatrix} w(k) \\ &+ f(k, x(k), \xi(k)) \end{aligned}$$

with the initial state $x(0)$ and the process noise $w(k)$ follows the zero-mean Gaussian distribution with the covariance $\Sigma_0 = \text{diag}\{5, 5\}$ and $Q(k) = 0.25$, respectively. The parameters of the dynamics of the sensors (9.2) are given as follows:

$$\begin{aligned} C_1(k) &= [0.92 + 0.05\cos(0.12k) \;\; 0.82] \\ C_2(k) &= [0.25 \;\; 0.1 + 0.05\sin(0.1k)] \\ C_3(k) &= [0.84 + 0.05\cos(0.1k) \;\; 0.75 + 0.05\sin(0.1k)] \\ C_4(k) &= [0.05 \;\; 0.05] \\ D_1(k) &= 0.18 \;\; D_2(k) = 0.12 \;\; D_3(k) = 0.16 \;\; D_4(k) = 0.14. \end{aligned}$$

Let the stochastic variables $\xi(k)$, $\zeta_i(k)$ and $v_i(k)$ be assumed to be uncorrelated zero-mean Gaussian white noise sequences with the covariance diag$\{1, 1\}$, diag$\{1, 1\}$, and $R_i(k) = 0.25$, $i = 1, 2, 3, 4$. The stochastic nonlinearities $f(k, x(k), \xi(k))$ and $g_i(k, x(k), \zeta_i(k))$ are selected as follows:

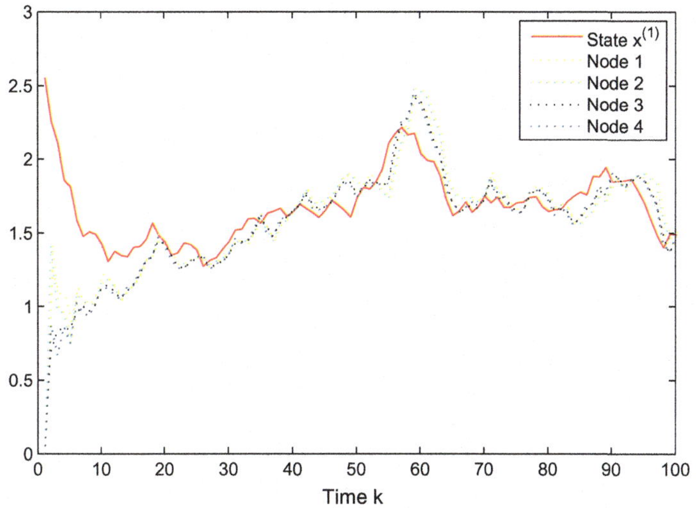

Fig. 9.1 The state x_1 and its estimation, $L = 1$

$$
\begin{aligned}
f(k, x(k), \xi(k)) &= \begin{bmatrix} 0.1 \\ 0.2 \end{bmatrix} (0.3\text{sign}(x^1(k))x^1(k)\xi^1(k) \\
&\quad + 0.4\text{sign}(x^2(k))x^2(k)\xi^2(k)) \\
g_i(k, x(k), \zeta_i(k)) &= 0.3(0.3\text{sign}(x^1(k))x^1(k)\zeta^1(k) \\
&\quad + 0.4\text{sign}(x^2(k))x^2(k)\zeta^2(k)),
\end{aligned}
$$

where $x^i(k)$, $\xi^i(k)$ and $\zeta^i(k)$ $(i = 1, 2)$ denote the ith elements of the system state $x(k)$, the stochastic variables $\xi(k)$ and $\zeta(k)$, respectively. Obviously, the first moment and the covariance of the nonlinearities meet the form of (9.3) and (9.4) with the integer $m = 1$, parameter matrices $\Pi_{s1} = [0.1\ 0.2]^T[0.1\ 0.2]$ and $\Pi_{s2} = 0.09$. Choose the positive scalar $\epsilon = 0.45$, which is smaller than $1/\Delta = 0.5556$. The initial local estimates are $\hat{x}_1(0) = \hat{x}_2(0) = \hat{x}_3(0) = \hat{x}_4(0) = [\ 0\ \ 0\ ]^T$. The designed filter gain matrices and the upper bound of estimation error covariance can be computed according to (9.14) and (9.15).

Figures 9.1, 9.2, 9.3, and 9.4 depict the system state and the tracking performance of consensus-based estimators with consensus step $L = 1$ and $L = 3$, respectively, from which we can find that the estimates of the four sensor nodes are roughly same when the consensus step increases. In Fig. 9.5, we compare the upper bound trace$\{\Xi_1(k)\}$ with the corresponding mean-square error (MSE) calculated by 300 Monte Carlo simulations. The result indicates that the solutions of the Riccati-like equation (9.15) are always the upper bounds of the MSEs.

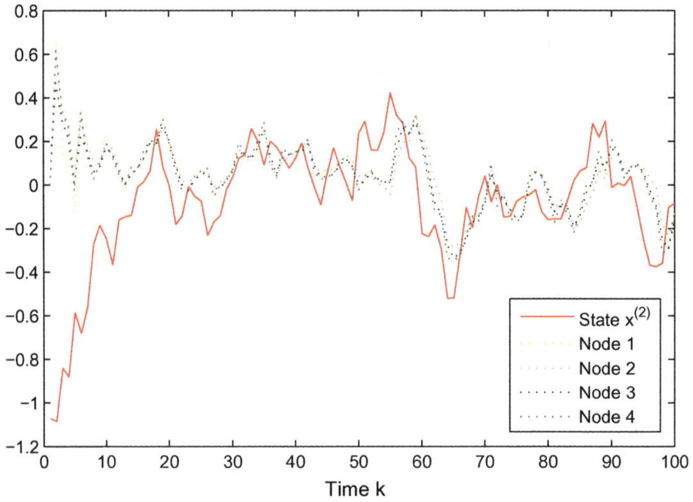

Fig. 9.2 The state x_2 and its estimation, $L = 1$

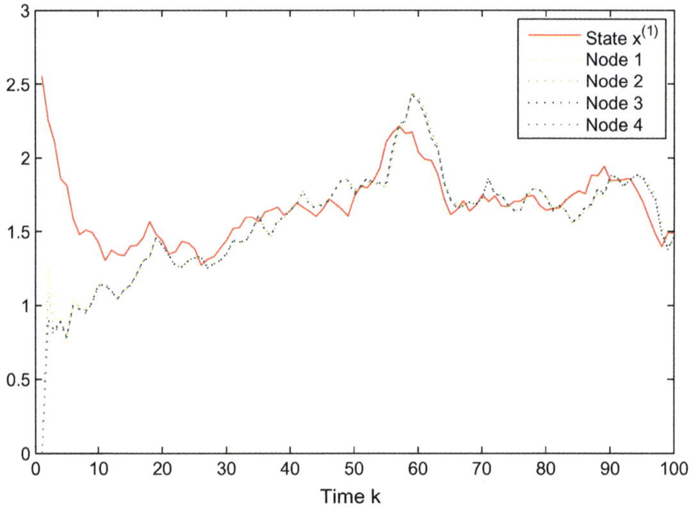

Fig. 9.3 The state x_1 and its estimation, $L = 3$

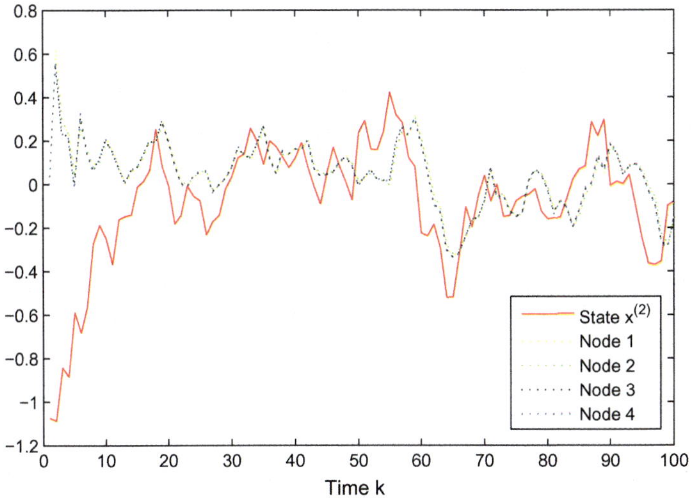

Fig. 9.4 The state x_2 and its estimation, $L = 3$

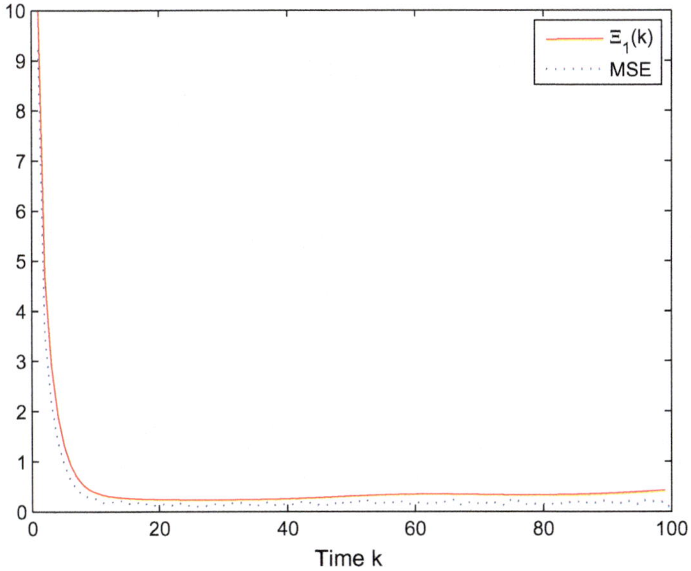

Fig. 9.5 MSE and its upper bound

9.4 Conclusions

In this article, the consensus-based filtering problem has been investigated for a class of discrete time-varying systems with stochastic nonlinearities. By employing the inductive method and the stochastic analysis technique, we have obtained Riccati-like equations, whose solutions are the upper bounds of the MSEs. Subsequently, the suboptimal filter gain matrices have been designed at every sampling instant in order to minimize the bounds. Finally, an illustrative example has been provided that validates the effectiveness of the designed consensus-based filters.

References

1. J. Yick, B. Mukherjee, D. Ghosal, Wireless sensor network survey. Comput. Netw. **52**(2008), 2292–2330 (2008)
2. D. Ding, Z. Wang, B. Shen, H. Shu, H_∞ state estimation for discrete-time complex networks with randomly occurring sensor saturations and randomly varying sensor delays. IEEE Trans. Neural Netw. **23**(5), 725–736 (2012)
3. H. Dong, Z. Wang, H. Gao, Distributed H_∞ filtering for a class of Markovian jump nonlinear time-delay systems over lossy sensor networks. IEEE Trans. Ind. Electron. **60**(10), 4665–4672 (2013)
4. R. Olfati-Saber, Kalman-consensus filter: optimality, stability, and performance, in *Proceedings of 48th IEEE Conference on Decision and Control*, Shanghai, P.R. China (2009), pp. 7036–7042
5. L. Xiao, S. Boyd, S. Lall, A scheme for robust distributed sensor fusion based on average consensus, in *Proceedings of IEEE Fourth International Symposium on Information Processing in Sensor, Networks* (2005), pp. 63–70
6. I. Matei, J. Baras, Consensus-based linear distributed filtering. Automatica **48**(8), 1776–1782 (2012)
7. E. Yaz, Y. Yaz, State estimation of uncertain nonlinear stochastic systems with general criteria. Appl. Math. Lett. **14**(5), 605–610 (2001)
8. J. Hu, Z. Wang, H. Gao, Recursive filtering with random parameter matrices, multiple fading measurements and correlated noises. Automatica **49**(2013), 3440–3448 (2013)
9. J. Hu, Z. Wang, H. Gao, L.K. Stergioulas, Extended Kalman filtering with stochastic nonlinearities and multiple missing measurements. Automatica **48**(9), 2007–2015 (2012)
10. W. Ren, R.W. Beard, Consensus seeking in multiagent systems under dynamically changing interaction topologies. IEEE Trans. Autom. Control **50**(5), 655–661 (2005)

Chapter 10
On Kalman-Consensus Filtering with Random Link Failures

Over the past decades, wireless sensor networks have attracted increasing research attention due primarily to their potential applications in various realms including seismic sensing, battlefield surveillance, intelligent transportation, and machine health monitoring [1]. A typical sensor network is composed of a group of autonomous sensor nodes spatially disseminated over certain monitored regions. Each sensor node shares the local information over the network via wireless communication to help the system complete complicated tasks in a cooperative manner.

As is well known, state estimation serves as one of the fundamental problems in the areas of signal processing and control engineering [2–7]. In particular, for the state estimation problems over sensor networks, the corresponding results can be categorized according to whether a fusion center exists or not. When a central node exists, the state estimation problem can be regarded as a centralized data fusion task whose aim is to obtain an improved estimate by merging all the measurements taken from multiple sources; see, e.g., [8, 9] and the reference therein. Although the centralized fusion strategy can achieve globally optimal results for the state estimation, a critical assumption is that all the observations (or the preprocessed data) shall be transmitted to the center. Unfortunately, due to inevitable cost constraints, such a centralized approach might be infeasible as it requires significant resource consumption. This is particularly true for sensor networks of large scale where the data communication constitutes the main source for energy usage. As such, an alternative approach called distributed state estimation has recently received much research attention, see, e.g., [10–15]. The main idea of the distributed algorithm is to decentralize the function of the fusion center by employing local estimators in every intelligent sensor, where each estimator only utilizes both the local information and the messages from the neighboring sensors (rather than all the sensors) to generate an estimate. When resource limit becomes a concern, such a distributed framework is preferable because the global knowledge of the network is no longer required by the individual estimators.

Q. Liu et al., *Stochastic Control and Filtering over Constrained Communication Networks*, Studies in Systems, Decision and Control 178,
https://doi.org/10.1007/978-3-030-00157-5_10

With the rapid technological developments of mobile sensor networks in recent years, the distributed estimation problems have started to attract research attention and a number of results have been reported in the literature. For example, the distributed version of the classical Kalman filter has been considered based on an innovation-exchanging technique in [16, 17]. Although utilizing the innovations from the neighboring nodes could help to improve the local estimates, there is no guarantee for the consensus of the estimates throughout the sensor networks. As such, several consensus-based distributed filtering algorithms, which share the common idea of disseminating the local information over the sensor network by performing a consensus procedure, have been proposed based on consensus on estimates [10], consensus on information matrices [18], and consensus on observations [19].

On the other hand, it has been well recognized that, among the inter-sensor communications in practical applications, temporary link failures happen frequently for a variety of reasons such as multipath fading, signal attenuation, background noise, external block, and so on [20, 21]. Much effort has been devoted to the investigation on how the link failures influence the distributed average over the connected notes. For instance, in [22], the authors have shown that the asymptotic agreement can be reached for the networks subject to an Erdős-Rényi random graph, where the communication behaviors between any pair of agents are governed by identically independent Bernoulli processes. As for a general class of random networks, it has been proved in [23] that the asymptotic properties can also be achieved if and only if the expected weights matrix is irreducible. Unfortunately, when it comes to the distributed estimation problems, the related results are scattered due probably to the lack of adequate analysis methods.

Motivated by the above discussion, in this chapter, we consider the distributed state estimation problem for a class of linear time-invariant systems over sensor networks subject to random communication failures. Following [24], a two-stage Kalman-consensus filtering algorithm is proposed that includes a classical Kalman-filter-based update as well as a consensus update of the information matrices/vectors. Different from [24], during the consensus procedure, the communication link is allowed to fail randomly with independent nonuniform probability. To the best of our knowledge, this chapter represents one of the first few attempts to handle the Kalman-consensus filter subject to link failures. The main contributions of this chapter are highlighted as follows. A sufficient condition is established for the stochastic boundedness of the filtering error by means of the collective observability and the network connectivity. By utilizing the ideas from the randomized consensus problems in [23], we prove that the uniform boundedness can be achieved almost surely for the filtering system when the number of consensus steps is sufficiently large.

This chapter is organized as follows. In Sect. 10.1, the sensor networks and the Kalman-consensus filtering algorithm are formally introduced. Some preliminary knowledge for boundedness analysis is presented in Sect. 10.2. The stochastic boundedness of the proposed Kalman-consensus filtering algorithm is investigated in Sect. 10.3. An illustrated example is presented in Sect. 10.4. Finally, Sect. 10.5 draws a conclusion about the chapter.

10.1 Problem Formulation

10.1.1 Topology Structure

An undirected graph $\mathcal{G} = (\mathcal{V}, \mathcal{E}, \mathcal{S})$ is utilized to describe the communication topology of the sensor network. Specifically, the graph $\mathcal{G}$ contains a vertex set $\mathcal{V} = \{v_1, v_2, \ldots, v_N\}$, an edge set $\mathcal{E} \subseteq \mathcal{V} \times \mathcal{V}$, and a weighted matrix $\mathcal{S} = [\pi_{ij}]_{N \times N}$, where N is the number of sensor nodes. The edge $(v_i, v_j) \in \mathcal{E}$ represents that the ith node can receive the messages from the jth node. The weighted matrix $\mathcal{S}$ is symmetric and doubly stochastic (i.e., each row and column sum to 1) with nonnegative elements π_{ij} satisfying the properties $\pi_{ij} > 0 \Longleftrightarrow (v_i, v_j) \in \mathcal{E}$. Moreover, since the nodes always have access to the local information, self-loops are permitted, i.e., $(v_i, v_i) \in \mathcal{E}$. The set of neighbors of the vertex v_i is denoted by $N_i \triangleq \{j : (v_i, v_j) \in \mathcal{E}\}$. A path of length l in an undirected graph $\mathcal{G}$ is a sequence of vertices $\{v_{i_1}, \ldots, v_{i_l}\}$ subject to $(v_{i_j}, v_{i_{j+1}}) \in \mathcal{E}$, for $1 \leq j \leq l - 1$. The length of the shortest path connecting the vertices v_i and v_j is called the distance between v_i and v_j. The maximum value of the distances between any two vertices in the graph $\mathcal{G}$ is called the diameter of $\mathcal{G}$ and is denoted by $\mathcal{D}(\mathcal{G})$. We assume that the graph $\mathcal{G}$ is connected throughout the chapter, i.e., for any pair of vertices $v_i, v_j \in \mathcal{V}$, there exists at least a path from v_i to v_j and vice versa.

10.1.2 Target Plant and Measurement Models

Consider a class of linear time-invariant systems described by the state-space model as follows:

$$x_{k+1} = Ax_k + w_k,$$

where $x_k \in \mathbb{R}^n$ is the current state of the process. The sensing model of the ith node is described by

$$y_{k,i} = C_i x_k + v_{k,i},$$

where $y_{k,i} \in \mathbb{R}^m$ is the measurement of the ith sensor at time instant k. The disturbances $w_k \in \mathbb{R}^n$ and $v_{k,i} \in \mathbb{R}^m$ are mutually independent white Gaussian random variables with zero-mean values and bounded covariances $Q > 0$ and $R_i > 0$, respectively. The initial state x_0 is a zero-mean Gaussian noise with the covariance $\Sigma_0 > 0$, and is uncorrelated with w_k and $v_{k,i}$ for all $k \geq 0$. The system parameters A and C_i are known matrices of appropriate dimensions. A mild assumption is made that the state transition matrix A is invertible.

10.1.3 Kalman-Consensus Filtering

First, a standard Kalman filter algorithm is applied to compute the optimal estimate for every individual sensor, which consists of the state prediction process

$$\begin{aligned} \hat{x}^i_{k|k-1} &= A\hat{x}^i_{k-1}, \\ P^i_{k|k-1} &= AP^i_{k-1}A' + Q, \end{aligned} \tag{10.1}$$

and the local measurement update process

$$\begin{aligned} \hat{x}^{i,o}_k &= \hat{x}^i_{k|k-1} + K^i_k(y_{k,i} - C_i\hat{x}^i_{k|k-1}), \\ P^{i,o}_k &= P_{k|k-1} - K^i_kC_iP^i_{k|k-1}, \end{aligned} \tag{10.2}$$

where $\hat{x}^i_{k|k-1}$ and $\hat{x}^{i,o}_k$ represent, respectively, the one-step prediction and updated estimate of the plant state with the corresponding error covariances $P^i_{k|k-1}$ and $P^{i,o}_k$. Moreover, the Kalman filter gain in (10.2) is given as follows:

$$K^i_k = P^i_{k|k-1}C'_i(C_iP^i_{k|k-1}C'_i + R_i)^{-1}.$$

In this chapter, we are interested in the situation where the individual sensor owns insufficient capabilities to ensure the local observability of a target plant. As a result, it is of particular importance that the sensor shall rely on its interconnections to make the state estimation problem feasible. On the other hand, it is well known that the estimates and covariance matrices from different sensors may contain complementary information. Combining the information from a variety of neighboring sources in a synergistic manner can help achieve a more accurate estimate of the plant state.

It should be pointed out that, in contrast to the settings in [8], the knowledge about the cross correlations between different sensors remains unknown in our chapter. Consequently, a multistep iterative covariance intersection, which is customarily referred to as the consensus procedure in this chapter, is adopted whose main idea is to combine the information matrices/vectors from neighboring sensors in a convex manner. Such a consensus procedure is implemented after the initial step of Kalman filtering in order to achieve the distributed state estimation, see [24] for more explanation. In this chapter, a consensus-based Kalman filtering algorithm (modified from that in [24] to allow random link failure) is outlined in Table 10.1. To avoid confusion in the following analysis, two different notations $P^{i,o}_k$ and P^i_k are utilized to represent the local covariances *before* and *after* the consensus update, respectively.

Table 10.1 Kalman-consensus Filtering Algorithm

Algorithm	
Step 1.	The sensors utilize the standard Kalman filter (10.1)–(10.2) to compute the locally optimal estimates.
Step 2.	Set the initial parameters $\Omega_k^i(0) = (P_k^{i,o})^{-1}$ and $q_k^i(0) = (P_k^{i,o})^{-1}\hat{x}_k^{i,o}$
Step 3.	Perform the consensus of L steps as follows: $\Omega_k^i(l+1) = \sum_{j \in N_i} \tilde{\pi}_{i,j}^{k,l} \Omega_k^j(l), \quad (10.3)$ $q_k^i(l+1) = \sum_{j \in N_i} \tilde{\pi}_{i,j}^{k,l} q_k^j(l).$
Step 4.	When the consensus is completed, calculate the renewed covariance and estimates by $P_k^i = [\Omega_k^i(L)]^{-1},\ \hat{x}_k^i = [\Omega_k^i(L)]^{-1} q_k^i(L). \quad (10.4)$

Note that the consensus update process is subject to certain degree of imperfect inter-sensor communication (e.g., random link failures). To facilitate the later analysis, the weights in (10.3) are defined as follows:

$$\tilde{\pi}_{i,j}^{k,l} = \begin{cases} \lambda_{i,j}^{k,l}\pi_{i,j}, & i \neq j \\ \pi_{i,i} + \sum\limits_{j \in N_i} (1 - \lambda_{i,j}^{k,l})\pi_{i,j}, & \text{otherwise}, \end{cases} \tag{10.5}$$

where $\left\{\lambda_{i,j}^{k,l} : k \geq 0, l \geq 1, j \geq i\right\}$ is a sequence of i.i.d. Bernoulli variables which govern the random link failures. Specifically, the communication between v_i and v_j ($(v_i, v_j) \in \mathcal{E}$) is successful if $\lambda_{i,j}^{k,l} = \lambda_{j,i}^{k,l} = 1$, otherwise it fails. For every communication link, we assume that it is always possible to deliver the data packets successfully, i.e.,

$$0 < \mathbb{P}\left(\lambda_{i,j}^{k,l} = 1\right) = \bar{\lambda}_{i,j} \leq 1,$$

for $i \neq j$, $k \geq 0$ and $0 \leq l \leq L-1$. Moreover, since the nodes have access to the local information regardless of the network conditions, we have $\lambda_{i,i}^{k,l} \equiv 1, \forall i \in \mathcal{V}$. The random variables $\lambda_{i,j}^{k,l}$ ($i \neq j$) are assumed to be independent with the process noise w_k, measurement noise v_k, and the initial state x_0. For ease of representation, a renewed weighted matrix $\tilde{\mathcal{S}}_{k,l} \triangleq [\tilde{\pi}_{i,j}^{k,l}]_{N \times N}$ is introduced to represent the random connections on the graph at the lth consensus step corresponding to instant k.

Remark 10.1 It can be seen from the Kalman-consensus filtering algorithm that the renewed weights $\tilde{\pi}_{i,j}^{k,l}$, $j \in N_i$ are based on the sequence $\lambda_{i,j}^{k,l}$, $j \in N_i$ whose exact knowledge is indispensable for the implementation of the consensus update. As such, sensors need to know whether they have received the messages from their neighboring nodes or not.

Remark 10.2 Various types of consensus weights $\pi_{i,j}^{k,l}$ have been utilized in the existing literature for different purposes. For example, [25] has utilized *Metropolis weights* such that the weighted matrix is doubly stochastic. To achieve the fastest convergence rate, [26] has designed the optimal consensus weights by solving a semidefinite program problem. In this chapter, since we need to ensure that the weighted matrix is doubly stochastic, a simple but effective way is to set the weights to be $\pi_{i,j}^{k,l} = 1/|N|$ for $(v_i, v_j) \in \mathcal{E}$.

10.2 Preliminary for Boundedness Analysis

In this section, some preliminary knowledge for the boundedness analysis of the proposed algorithm is presented. We will reveal some properties of the generalized weighted matrix and show the consistency of the Kalman-consensus filter.

Some useful lemmas are listed as follows:

Lemma 10.2.1 *Given positive semidefinite matrices A and B, if $A + B > 0$, then the inequality $\alpha A + \beta B > 0$ always holds for any positive scalars α and β.*

Proof Since A and B are positive semidefinite, and α and β are positive, one has $\alpha A + \beta B \geq 0$. Next, we will prove that $\alpha A + \beta B$ has no zero eigenvalue. Assume that there exists a nonzero vector x such that $(\alpha A + \beta B)x = 0$. Multiplying x' on both sides of the equation yields $x'(\alpha A + \beta B)x = 0$, which implies $x \in \mathcal{N}(A) \cap \mathcal{N}(B)$. Obviously, we have $(A + B)x = 0$ and that contradicts the condition $A + B > 0$. Therefore, it is concluded that $\alpha A + \beta B > 0$.

Lemma 10.2.2 *Given a deterministic graph $\mathcal{G}$, there exists a path no longer than l that starts from v_i and ends at v_j if and only if the (i, j)th entry of the matrix $\mathcal{S}^l$ is nonzero.*

Proof First, let us use the inductive method to show the sufficiency. When $l = 1$, it is clear that $\mathcal{S}^1 = \mathcal{S}$, so the statement is apparently true. Assume that the result is true when $l = t$. For $l = t + 1$, we have

$$[\mathcal{S}^{t+1}]_{ij} = \sum_{s=1}^{N} [\mathcal{S}^t]_{is} \pi_{sj},$$

where $[\mathcal{S}^t]_{is}$ is the (i, s)th entry of $\mathcal{S}^t$. Since $[\mathcal{S}^{t+1}]_{ij} > 0$, it is not hard to verify that there exists a nonzero pair $([\mathcal{S}^t]_{is'}, \pi_{s'j})$. Therefore, v_i can reach v_j via the critical vertex $v_{s'}$ with the path no longer than $t + 1$, which ends the proof. The necessity can be proved in a similar way and thus is omitted here.

As is well known, the weighted matrix $\mathcal{S}$ shows the one-step connections between any two vertices in the graph, i.e., $\pi_{ij} > 0$ if and only if $(v_i, v_j) \in \mathcal{E}$. Lemma 10.2.2 further reveals that the matrix $\mathcal{S}^l$ represents the generalized connections between two vertices in the graph via a path no longer than l. From this point, it can be seen that, when the power l is no less than the diameter of the graph $\mathcal{G}$ (i.e., $l \geq \mathcal{D}(\mathcal{G})$), every entry in the matrix $\mathcal{S}^l$ shall be positive.

Next, we will show that the Kalman-consensus filtering algorithm proposed in this chapter yields a *consistent* fused estimate. The phrase "consistent" here indicates that the actual covariance is always bounded by the estimated covariance. To this end, we first denote the actual one-step prediction error covariance by

$$P^i_{act,k|k-1} \triangleq \mathbb{E}[(x_k - \hat{x}^i_{k|k-1})(x_k - \hat{x}^i_{k|k-1})'],$$

the actual estimation error covariance by

$$P^{i,o}_{act,k} \triangleq \mathbb{E}[(x_k - \hat{x}^{i,o}_k)(x_k - \hat{x}^{i,o}_k)'],$$

and the actual estimation error covariance after multistep consensus update by

$$P^i_{act,k} \triangleq \mathbb{E}[(x_k - \hat{x}^i_k)(x_k - \hat{x}^i_k)'].$$

Then, a proposition can be obtained as follows.

Proposition 10.2.3 *The Kalman-consensus filtering algorithm subject to stochastic link failures is individually consistent, i.e.,* $P^i_{act,k} \leq P^i_k$, $P^i_{act,k+1|k} \leq P^i_{k+1|k}$, $\forall i = 1, \ldots, N$.

Proof If $P^i_{act,k} \leq P^i_k$ is true, then the conclusion $P^i_{act,k+1|k} \leq P^i_{k+1|k}$ can be derived directly. Therefore, in the following proof, we will demonstrate that the inequality $P^i_{act,k} \leq P^i_k$ holds by using an inductive method. When $k = 0$, the result is clearly true. Suppose now that, when $k = t$, the consistency of the individual estimates is satisfied (i.e., $P^i_{act,t} \leq P^i_t$, for $i = 1, 2, \ldots, N$). We introduce an augmented matrix as follows:

$$\mathbf{\Omega_k}(l) = \text{col}\{\Omega^1_k(l), \ldots, \Omega^N_k(l)\}.$$

Running L step consensus updates according to *Step 3* in the Kalman-consensus filtering algorithm yields

$$\begin{aligned} \mathbf{\Omega_k}(L) &= \underbrace{(\tilde{\mathcal{S}}_{k,L-1} \otimes I) \cdots (\tilde{\mathcal{S}}_{k,0} \otimes I)}_{L} \mathbf{\Omega_k}(0) \\ &= (\tilde{\mathcal{S}}^L_k \otimes I)\mathbf{\Omega_k}(0), \end{aligned} \tag{10.6}$$

where

$$\tilde{\mathcal{S}}_k^L := \prod_{p=1}^{L} \tilde{\mathcal{S}}_{k,L-p}.$$

Noting that $(P_k^i)^{-1} = \Omega_k^i(L)$ is the ith row matrix block in the augmented matrix $\mathbf{\Omega_k}(L)$, we have the following equation at the time instant $k = t + 1$:

$$\left(P_{t+1}^i\right)^{-1} = \sum_{j=1}^{N} \tilde{\pi}_{i,j}^L \left(P_{t+1}^{j,o}\right)^{-1}, \tag{10.7}$$

where $\tilde{\pi}_{i,j}^L$ is the (i, j)th entry of the matrix $\tilde{\mathcal{S}}_k^L$. Similarly, it can be verified that

$$\left(P_{t+1}^i\right)^{-1} \hat{x}_k^i = \sum_{j=1}^{N} \tilde{\pi}_{i,j}^L \left(P_{t+1}^{j,o}\right)^{-1} \hat{x}_k^{j,o}. \tag{10.8}$$

Since $\tilde{\mathcal{S}}_k^L$ is a product of a set of row stochastic matrices, it is still a row stochastic matrix, i.e., $\sum_{j=1}^{N} \tilde{\pi}_{i,j}^L = 1$ and $\tilde{\pi}_{i,j}^L \geq 0$. From the fact that $P_{act,t}^i \leq P_t^i$ and the properties of Riccati equations concerning the Kalman filter, we have $P_{act,t+1}^{i,o} \leq P_{t+1}^{i,o}$. According to the results in Section II of the work [27], it is straightforward to verify that $P_{act,t+1}^i \leq P_{t+1}^i$, which concludes the proof.

From above proposition, we can see that, for arbitrary cross correlations, the proposed Kalman-consensus filtering algorithm still yields the estimate consistency, which is essential for establishing the conditions for stochastic boundedness.

10.3 Boundedness Analysis

In this section, we are dedicated to the development of a sufficient condition for the stochastic boundedness of the proposed Kalman-consensus filtering scheme subject to random link failures. The situations involving the finite and infinite number of consensus steps are discussed, respectively.

We first present the concept of collective observability as follows.

Assumption 10.1 (*Collective Observability*) The sensor network is collectively observable, i.e., (A, C) is observable, where $C \triangleq [C_1', C_2', \ldots, C_N']'$.

Owing to the constraints from sensor networks in practice, it would be restrictive and even unreasonable to assume the local observability which often refers to the observability of the target from the observations taken by every node (e.g., (A, C_i), $\forall i \in \mathcal{V}$ is observable). By collective observability, we mean that the observability is now achieved by the sensor network *as a whole*.

Remark 10.3 The collective observability in Assumption 10.1 is equivalent to the statement that there exists $l > 0$ such that the Grammian matrix satisfies $\sum_{k=1}^{l}(A^{k-1})'C'CA^{k-1} > 0$. Furthermore, noting that the covariances of the measurement noises R_i are bounded and positive definite, we can conclude that, if the sensor network is collectively observable, then $\sum_{k=1}^{l}\sum_{i=1}^{N}(A^{k-l})'C_i'(R_i)^{-1}C_iA^{k-l} > 0$ holds.

10.3.1 Finite Number of Consensus Steps

In this case, the finite number of consensus steps is under consideration. Before proceeding, inspired by [24], the uniform boundedness of the matrix P_k^i can be shown in the following proposition.

Proposition 10.3.1 *Under Assumptions 10.1, there exists an instant p and a positive scalar* α_i*, such that when no communication failure occurs during the interval* $[k, k+p]$ *(i.e.,* $\lambda_{i,j}^{k,l} = 1$*) the following uniform boundedness condition*

$$P_{k+p}^i \leq \alpha_i I, \quad \forall P_k^1, \ldots, P_k^N \in \mathbb{S}_+^n$$

is true for $i = 1, 2, \ldots, N$.

Proof According to the propagation of the one-step predict error covariance in (10.1), we have

$$\begin{aligned}(P_{k|k-1}^i)^{-1} &= (AP_{k-1}^iA' + Q)^{-1} \\ &= (A^{-1})'(P_{k-1}^i + A^{-1}Q(A^{-1})')^{-1}A^{-1}.\end{aligned} \tag{10.9}$$

Obviously, there always exists a positive scalar $\beta > 0$ such that the following condition holds:

$$A^{-1}Q(A^{-1})' \leq \beta P_{k-1}^i.$$

By letting $\gamma = (1+\beta)^{-1}$ and substituting the above inequality into (10.9), one has

$$(P_{k|k-1}^i)^{-1} \geq \gamma(A^{-1})'(P_{k-1}^i)^{-1}A^{-1}. \tag{10.10}$$

Running L step consensus updates according to *Step 3* in the Kalman-consensus filtering algorithm yields

$$\begin{aligned}\mathbf{\Omega_k}(L) &= \underbrace{(\mathcal{S} \otimes I)\cdots(\mathcal{S} \otimes I)}_{L}\,\mathbf{\Omega_k}(0) \\ &= (\mathcal{S}^L \otimes I)\mathbf{\Omega_k}(0).\end{aligned}$$

For ease of representation, we let $\pi_{i,j}^L$ be the (i, j)th entry of the matrix $\mathcal{S}^L$ and obtain

$$\begin{aligned}(P_k^i)^{-1} &= \sum_{j=1}^{N} \pi_{i,j}^L (P_k^{j,o})^{-1} \\ &= \sum_{j=1}^{N} \pi_{i,j}^L (P_{k|k-1}^j)^{-1} + \sum_{j=1}^{N} \pi_{i,j}^L C_j'(R_j)^{-1} C_j,\end{aligned}$$

where the second equality comes from the measurement update process of the local Kalman filter (10.2). Substituting (10.10) into the above equation yields

$$\begin{aligned}&(P_{k+p}^i)^{-1} \\ &\geq \gamma \sum_{j=1}^{N} \pi_{i,j}^L (A^{-1})'(P_{k+p-1}^j)^{-1} A^{-1} + \sum_{j=1}^{N} \pi_{i,j}^L C_j'(R_j)^{-1} C_j.\end{aligned}$$

Once again, we utilize the above inequality with respect to $(P_{k+p}^j)^{-1}$ and have

$$\begin{aligned}&(P_{k+p}^i)^{-1} \\ &\geq \gamma \sum_{j=1}^{N} \pi_{i,j}^L (A^{-1})' \left[\gamma \sum_{s=1}^{N} \pi_{j,s}^L (A^{-1})'(P_{k+p-2}^s)^{-1} A^{-1} \right. \\ &\quad \left. + \sum_{s=1}^{N} \pi_{j,s}^L C_s'(R_s)^{-1} C_s \right] A^{-1} + \sum_{j=1}^{N} \pi_{i,j}^L C_j'(R_j)^{-1} C_j \\ &\geq \gamma^2 \sum_{j=1}^{N} \sum_{s=1}^{N} \pi_{i,j}^L \pi_{j,s}^L (A^{-2})'(P_{k+p-2}^s)^{-1} A^{-2} \\ &\quad + \gamma \sum_{j=1}^{N} \pi_{i,j}^L (A^{-1})' \sum_{s=1}^{N} \pi_{j,s}^L C_s'(R_s)^{-1} C_s A^{-1} \\ &\quad + \sum_{j=1}^{N} \pi_{i,j}^L C_j'(R_j)^{-1} C_j. \end{aligned} \tag{10.11}$$

Noting that $\mathcal{S}^{2L} = \mathcal{S}^L \mathcal{S}^L$, we have $\pi_{i,s}^{2L} = \sum_{j=1}^{N} \pi_{i,j}^L \pi_{j,s}^L$ and therefore

$$\begin{aligned}
&\sum_{j=1}^{N}\sum_{s=1}^{N}\pi_{i,j}^{L}\pi_{j,s}^{L}(P_{k+p-2}^{s})^{-1}\\
&=\sum_{s=1}^{N}\sum_{j=1}^{N}\pi_{i,j}^{L}\pi_{j,s}^{L}(P_{k+p-2}^{s})^{-1}\\
&=\sum_{s=1}^{N}\pi_{i,s}^{2L}(P_{k+p-2}^{s})^{-1}. \qquad (10.12)
\end{aligned}$$

Using the recursion in (10.11) and employing (10.12), we arrive at

$$\begin{aligned}
&(P_{k+p}^{i})^{-1}\\
&\geq \gamma^{p}\sum_{j=1}^{N}\pi_{i,j}^{pL}(A^{-p})'(P_{k}^{j})^{-1}A^{-p}\\
&+\sum_{j=1}^{N}\sum_{l=1}^{p}\gamma^{(p-l)}\pi_{i,j}^{(p-l+1)L}(A^{l-p})'C_{j}'(R_{j})^{-1}C_{j}A^{l-p},
\end{aligned}$$

which results in

$$(P_{k+p}^{i})^{-1}\geq\sum_{j=1}^{N}\sum_{l=1}^{p}h_{i,j,l}(A^{l-p})'C_{j}'(R_{j})^{-1}C_{j}A^{l-p}, \qquad (10.13)$$

where $h_{i,j,l}=\gamma^{(p-l)}\pi_{i,j}^{(p-l+1)L}$. Choosing $p>\mathcal{D}(\mathcal{G})/L+n$, according to Lemma 10.2.2, we have $\pi_{i,j}^{(p-n)L}>0$. Obviously,

$$\begin{aligned}
&(P_{k+p}^{i})^{-1}\\
&\geq\sum_{j=1}^{N}\sum_{l=p-n}^{p}\gamma^{(p-l)}\pi_{i,j}^{(p-l+1)L}(A^{l-p})'C_{j}'(R_{j})^{-1}C_{j}A^{l-p}.
\end{aligned}$$

Note that now the right-hand side of the above inequality is independent of the error covariance. In view of Remark 10.3, there must exist a positive scalar α_i such that the updated error covariance is uniformly bounded at the end of the interval $[k, k+p]$, i.e., $P_{k+p}^{i}\leq\alpha_i I$. The proof is complete.

Up to now, we have presented the sufficient criterion for the boundedness of the covariance under the perfect communication condition. In the situation where sensors suffer from link failures, the stochastic boundedness will be proved in the following analysis. For presentation convenience, we denote

$$\Theta_k \triangleq \left\{\Lambda_{k,l} = \left[\lambda_{i,j}^{k,l}\right]_{N\times N}, 0 \leq l \leq L-1\right\},$$

where $\Lambda_{k,l} \in \mathbb{R}^{N\times N}$ is an indicator variable whose (i, j)th entry $\lambda_{i,j}^{k,l}$ represents whether the edge (v_i, v_j) suffers from a communication link failure at the instant k. Moreover, a set describing the communication statuses during the interval $[k-p+1, k]$ is denoted as follows:

$$\mathfrak{S}_k \triangleq \left\{\Theta_k, \Theta_{k-1}, \ldots, \Theta_{k-p+1}\right\}.$$

Noting that the number of state variables in the set $\mathfrak{S}_k$ is finite, we introduce a finite set Ψ that contains all possible variables of $\mathfrak{S}_k$, i.e., $\mathfrak{S}_k \in \Psi$. Furthermore, we denote a set to represent the event that no communication failure occurs by

$$\tilde{\mathfrak{S}} \triangleq \left\{\tilde{\Theta}_1, \tilde{\Theta}_2, \ldots, \tilde{\Theta}_p\right\},$$

where

$$\tilde{\Theta}_k \triangleq \left\{\tilde{\Lambda}_{k,l} = [1]_{N\times N}, 0 \leq l \leq L-1\right\}.$$

Obviously, it follows that $\tilde{\mathfrak{S}} \in \Psi$.

By exploiting the stochastic analysis technique, we have the following result.

Theorem 10.3.2 *Under Assumptions 10.1, the error covariance of the sensor network is stochastically bounded, i.e.,*

$$\lim_{\epsilon\to\infty} \sup_{k\in\mathbb{N}} \mathbb{P}\left(\left\|\sum_{i=1}^{N} P_{act,k|k-1}^{i}\right\| > \epsilon\right) = 0.$$

Proof It has been proved in (10.10) that

$$(P_{k+1|k}^{i})^{-1} \geq \gamma(A^{-1})'(P_k^i)^{-1}A^{-1}.$$

Considering the consensus update process, one has

$$(P_{k+1|k}^{i})^{-1} \geq \gamma(A^{-1})'\sum_{j=1}^{N}\tilde{\pi}_{i,j}^{L}(P_k^{j,o})^{-1}A^{-1}.$$

Since the sensors always have access to the local information (i.e., $\lambda_{i,i}^{k,l} = 1$ and $\pi_{i,i} > 0, \forall i \in \mathcal{V}$), it is obvious that $\tilde{\pi}_{i,i}^{L} > 0$, which implies that a lower bound $r > 0$ of $\tilde{\pi}_{i,i}^{L}$ exists, i.e., $\tilde{\pi}_{i,i}^{L} > r$. According to the properties of the Kalman filter algorithm, we have $(P_k^{i,o})^{-1} \geq (P_{k|k-1}^i)^{-1}$. Combining such an inequality with the fact $\tilde{\pi}_{i,j}^L \geq 0$ and $\tilde{\pi}_{i,i}^L > 0$, we see that there exists a positive scalar $\tilde{\gamma}$ such that the following inequality holds:

$$(P_{k+1|k}^i)^{-1} \geq \tilde{\gamma}(A^{-1})'(P_{k|k-1}^i)^{-1}A^{-1}. \tag{10.14}$$

For the convenience of representation, we define $\mathcal{M}_{k+1} \triangleq P_{k+1|k}^i$ and it is clear that there exists a positive scalar $\mu > 0$ such that the following inequality holds:

$$\mathcal{M}_{k+1} < \tilde{\gamma}^{-1}A'\mathcal{M}_k A + \mu I. \tag{10.15}$$

Let

$$\kappa(\epsilon) = \max_{k\in\mathbb{N}} \left\{ k \in \mathbb{N} \left| \tilde{\alpha}\zeta^{2k} + \frac{\zeta^{2k} - 1}{\zeta^2 - 1}\mu \leq \epsilon \right. \right\}, \tag{10.16}$$

where $\tilde{\alpha} = \|A\|^2\alpha_i + \|Q\|$, and $\zeta = \tilde{\gamma}^{-1/2}\|A\|$. By choosing a small $\tilde{\gamma}$, we can guarantee $\zeta > 1$ subject to (10.14). Consequently, it is not hard to verify that $\kappa(\epsilon) \to \infty$ when $\epsilon \to \infty$.

Next, we will prove the following inequality holds:

$$\begin{aligned} &\mathbb{P}\left(\left\|\mathcal{M}_{k_0+1}\right\| > \epsilon\right) \\ &\quad \leq \mathbb{P}\left(\text{no event } \tilde{\mathfrak{S}} \text{ occurs in } k \in [k_0 - \kappa(\epsilon), k_0]\right). \end{aligned} \tag{10.17}$$

Suppose that the event $\tilde{\mathfrak{S}}$ does occur during the interval $[k_0 - \kappa(\epsilon), k_0]$. Therefore, there exists an instant $k' \in [k_0 - \kappa(\epsilon), k_0]$ such that $\Theta_{k'-p+l} = \tilde{\Theta}_l, \forall l = 1, 2, \ldots, p$. According to Proposition 10.3.1, it can be seen that $P_{k'}^i \leq \alpha_i I$ and then it follows from the prediction equation (10.1) that

$$\mathcal{M}_{k'+1} = AP_{k'}^i A' + Q \leq \alpha_i\|A\|^2 I + \|Q\|I = \tilde{\alpha}I. \tag{10.18}$$

Taking norms to both sides of (10.15) yields

$$\|\mathcal{M}_{k+1}\| \leq \tilde{\gamma}^{-1}\|A\|^2\|\mathcal{M}_k\| + \mu.$$

Accordingly, it follows from the above inequality and (10.18) that

$$\begin{aligned} \|\mathcal{M}_{k_0+1}\| &\leq \tilde{\alpha}\zeta^{2(k_0-k')} + \frac{\zeta^{2(k_0-k')} - 1}{\zeta^2 - 1}\mu \\ &\leq \tilde{\alpha}\zeta^{2\kappa(\epsilon)} + \frac{\zeta^{2\kappa(\epsilon)} - 1}{\zeta^2 - 1}\mu \leq \epsilon, \end{aligned}$$

where the fact that $k_0 - k' \leq \kappa(\epsilon)$ has been utilized to derive the first inequality and the definition of $\kappa(\epsilon)$ in (10.16) has been exploited in reaching the second inequality. So far, it can be inferred that, when the event $\tilde{\mathfrak{S}}$ occurs in the interval $[k_0 - \kappa(\epsilon), k_0]$, the inequality $\|\mathcal{M}_{k_0+1}\| \leq \epsilon$ is always true, which further yields the following inclusion:

$$\Big\{\text{the event } \tilde{\mathfrak{S}} \text{ occurs in } k \in [k_0 - \kappa(\epsilon), k_0]\Big\} \subset \Big\{\|\mathcal{M}_{k_0+1}\| \leq \epsilon\Big\}$$

Therefore, the inequality (10.17) holds. To continue the analysis, we define two instants k_1 and k_2 as follows:

$$k_1 \triangleq \min_k\{k \geq p | \mathfrak{S}_k = \tilde{\mathfrak{S}}\},$$
$$k_2 \triangleq \min_k\{k \geq k_0 - \kappa(\epsilon) | \mathfrak{S}_k = \tilde{\mathfrak{S}}\}.$$

From the total probability formula and the inequality (10.17), it follows that

$$\begin{aligned}\mathbb{P}\left(\left\|\mathcal{M}_{k_0+1}\right\| > \epsilon\right) &\leq \mathbb{P}(k_2 > k_0) \\ &= \sum_{\varpi \in \Psi} \mathbb{P}(k_2 > k_0 | \mathfrak{S}_{k_0-\kappa(\epsilon)-1} = \varpi)\mathbb{P}(\mathfrak{S}_{k_0-\kappa(\epsilon)-1} = \varpi).\end{aligned} \tag{10.19}$$

Obviously, the following relationship is true:

$$\mathbb{P}(k_2 > k_0 | \mathfrak{S}_{k_0-\kappa(\epsilon)-1} = \varpi) = \mathbb{P}(k_1 > \kappa(\epsilon) + 1 | \mathfrak{S}_p = \varpi). \tag{10.20}$$

Based on the definition of $\kappa(\epsilon)$ in (10.16), we have $\kappa(\epsilon) \to \infty$ as $\epsilon \to \infty$ and, furthermore,

$$\lim_{\epsilon \to \infty} \mathbb{P}(k_1 > \kappa(\epsilon) + 1 | \mathfrak{S}_p = \varpi) = 0. \tag{10.21}$$

Combining (10.17) with (10.19)–(10.21) and noting that the set Ψ only contains finite states, we arrive at $\lim_{\epsilon\to\infty} \sup_{k\in\mathbb{N}} \mathbb{P}\left(\|\mathcal{M}_{k+1}\| > \epsilon\right) = 0$. According to Proposition 10.2.3, we obtain the following inclusion:

$$\left\{\left\|P^i_{act,k|k-1}\right\| > \epsilon\right\} \subseteq \left\{\left\|P^i_{k|k-1}\right\| > \epsilon\right\},$$

which further leads to the following conclusion:

$$\lim_{\epsilon\to\infty} \sup_{k\in\mathbb{N}} \mathbb{P}\left(\left\|\sum_{i=1}^{N} P^i_{act,k|k-1}\right\| > \epsilon\right) = 0.$$

The proof is now complete.

Remark 10.4 In contrast to [28], any failure probability less than 1 can guarantee the stochastic boundedness of the covariance in our chapter. An important reason is that the stochastic boundedness is weaker than the mean boundedness considered in [28]. That is, if the mean covariance is bounded then the stochastic boundedness holds, but not vice versa.

Remark 10.5 As mentioned previously, the collective observability is essential to the state estimation problem in sensor networks. Such a concept has been initially introduced in sensor fusion problems so as to make sure that the center estimator can generate a satisfactory estimate based on the measurements taken from all the sensors. Different from traditional literature, in this chapter, the distributed estimation problem is considered where the local information can be only disseminated between spatially close nodes. It should be pointed out that the collective observability itself is now *not* sufficient to guarantee the boundedness of the filtering errors since the topology structure has an impact on the information flows in the network. From the above proofs, it can be seen that a connected graph could help the local information to spread throughout the network, thereby making the distributed state estimation problem feasible.

10.3.2 Infinite Number of Consensus Steps

In the previous subsection, the stochastic boundedness of the Kalman-consensus filtering algorithm under finite number of consensus steps is investigated. Now, let us turn to another situation where a sufficiently large number of steps for the consensus are permitted. First, a lemma on the consensus over ergodic stationary graph processes is presented as follows.

Lemma 10.3.3 ([23]) *Let $\{\tilde{\mathcal{S}}_{k,l}, l, k \geq 0\}$ denote an ergodic stationary process of stochastic matrices with strictly positive diagonals.*

$$\lim_{l\to\infty} \tilde{\mathcal{S}}_{k,l}\cdots\tilde{\mathcal{S}}_{k,2}\tilde{\mathcal{S}}_{k,1} = \mathbf{1}d'$$

holds almost surely if and only if $|\lambda_2(\mathbb{E}[\tilde{\mathcal{S}}_{k,l}])| < 1$, where $d = [d_1, \ldots, d_N]' \in \mathbb{R}^N$ is a random vector satisfying $d'\mathbf{1} = 1$.

The above lemma implies that the almost sure asymptotic consensus over ergodic stationary graph processes can be achieved if and only if the expected stochastic matrix is irreducible. Bearing this in mind, we have the following theorem.

Theorem 10.3.4 *Under Assumption 10.1, if the number of consensus steps is sufficiently large, there must exist positive scalars $\check{\alpha}_i$ such that that the following uniformly bounded conditions*

$$\sup_{k\in\mathbb{N}} P^i_{act,k} \leq \breve{\alpha}_i I, \ \forall P^1_0, \ldots, P^N_0 \in \mathbb{S}^n_+,$$

are satisfied almost surely for $i = 1, 2, \ldots, N$.

Proof According to the Kalman-consensus filtering algorithm, we have

$$\boldsymbol{\Omega}_{\mathbf{k}}(L) = (\tilde{\mathcal{S}}^L_k \otimes I)\Omega_{\mathbf{k}}(0).$$

where $\tilde{\mathcal{S}}^L_k \triangleq \prod_{p=1}^L \tilde{\mathcal{S}}_{k,L-p}$. Since $\mathbb{P}(\tilde{\pi}^{k,l}_{i,j} > 0) > 0$, for $j \in N_i, i \in \mathcal{V}, k \geq 0, 0 \leq l \leq L-1$, it is clear that the expectation of the random matrix $\tilde{\mathcal{S}}_{k,l}$ has the same structure as $\mathcal{S}$ and is also an irreducible matrix implying $|\lambda_2(\mathbb{E}[\tilde{\mathcal{S}}_{k,l}])| < 1$. Moreover, from (10.5), it can be seen that $\tilde{\mathcal{S}}_{k,l}$ is doubly stochastic. In the light of Lemma 10.3.3 and Theorem 4 in [23], the limit $\tilde{\mathcal{S}}_{k,L} \cdots \tilde{\mathcal{S}}_{k,2}\tilde{\mathcal{S}}_{k,1} \to \mathbf{1}d'$ where $d = (1/N)\mathbf{1}$, holds almost surely. When $1 \leq k \leq p-1$, we can always find a positive scalar $\breve{\beta}_{i,1}$ during a finite interval such that $P^i_k \leq \breve{\beta}_{i,1} I$ is satisfied. As for the case $k \geq p$, similar to the proof of Proposition 10.3.1, we can verify that

$$\begin{aligned}&(P^i_{k+p})^{-1}\\ &\geq \gamma^{p+1} \sum_{j=1}^N \sum_{l=1}^p \gamma^{(p-l)} \tilde{\pi}^\infty_{i,j} (A^{l-p})' C'_j (R_j)^{-1} C_j A^{l-p}.\end{aligned}$$

Here, $\tilde{\pi}^\infty_{i,j}$ is the (i, j)th entry of the matrix $\tilde{\mathcal{S}}_{k,L} \cdots \tilde{\mathcal{S}}_{k,2}\tilde{\mathcal{S}}_{k,1}$. Since d is nonnegative, when the number of consensus steps is infinite, $\tilde{\pi}^\infty_{i,j} > 0$ holds almost surely. Once again, by utilizing the collective observability condition, there exists a positive scalar $\breve{\beta}_{i,2}$ such that $P^i_k \leq \breve{\beta}_{i,2} I$. Choosing $\breve{\alpha}_i = \max\{\breve{\beta}_{i,1}, \breve{\beta}_{i,2}\}$, we can see that $P^i_k \leq \breve{\alpha}_i I$. To this end, according to Proposition 10.2.3, we can conclude that $P^i_{act,k} \leq \breve{\alpha}_i I$ and the proof is now complete.

Remark 10.6 In this chapter, we consider the distributed state estimation problem over wireless sensor networks with unreliable communication links. A Kalman-consensus filtering approach is developed that possesses the consistent property, that is, the estimated error covariance is always an upper bound of the actual error covariance. The main novelties lie in two aspects: (1) the stochastic boundedness of the filtering error dynamics is investigated by means of the collective observability as well as the network connectivity; and (2) the uniform boundedness is proven to be achievable almost surely for the filtering system when the number of consensus steps is sufficiently large.

Remark 10.7 In [29], the distributed Kalman filtering problems have been studied and some elegant results have been obtained. Nevertheless, the objective of [29] is to design the optimal filter gain by utilizing the statistical properties of the random link failures, whereas our chapter focuses on the boundedness analysis of the Kalman-consensus filter. Moreover, compared with [30], the link failures are further taken into consideration in our chapter. Note that the link failures have a great impact on the filtering performance and sometime might even diverge the estimation error covariance, so we need to investigate when the covariance is stochastically bounded.

10.4 Simulation Example

In this section, a numerical example is presented to verify the results established in this chapter.

Consider a target tracking system described by a linear model as follows:

$$x_{k+1} = \begin{bmatrix} 1 & dt & 0 & 0 \\ 0 & 1.08 & 0 & 0 \\ 0 & 0 & 1 & dt \\ 0 & 0 & 0 & 1.08 \end{bmatrix} x_k + w_k,$$

where $dt = 0.01$ is the discretization step size and w_k is a Gaussian disturbance with the covariance $Q = dtI$. The initial state x_0 is assumed to obey the Gaussian distribution with zero mean and covariance $4I$. In this scenario, a group of sensors ($n = 70$) are deployed to track the target state x_k, 10 of which are able to measure the position on the x-axis, 10 of which are able to measure the position on the y-axis, whereas the remaining nodes only have the communication and signal processing capabilities. The considered scenario is presented in Fig. 10.1. Moreover, the sensor shares the local information with other sensors within its communication radius 5. For sake of convenience, the weights are set to be $\pi_{ij} = 1/|N_i|$ for $(v_i, v_j) \in \mathcal{E}$, where $|N_i|$ is the number of the neighboring nodes of sensor i. Moreover, the probabilities of successful transmissions are given as follows:

$$\mathbb{P}\left(\lambda_{i,j}^{k,l} = 1\right) = 0.9, \ (v_i, v_j) \in \mathcal{E}.$$

The sensors measure the position of the target according to the following observation model:

$$y_{k,i} = C_i x_k + v_{k,i}$$

where v_k is an additive Gaussian disturbance with zero mean and the covariance $R_i = I$. The sensor nodes provide measurements of the position of the target plant in Cartesian coordinates. The measurement matrices are chosen to be

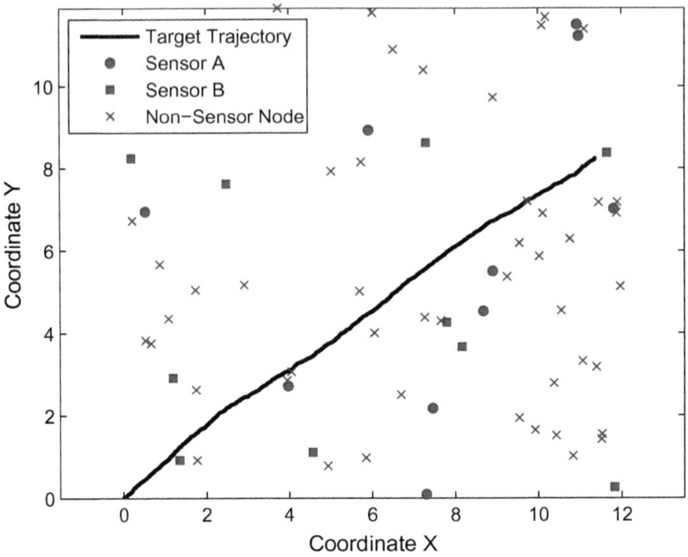

Fig. 10.1 A wireless sensor network. Sensor A and Sensor B are able to measure the position of the target on the x-axis and y-axis, respectively. Non-Sensor Node only has the communication and signal processing capabilities

$$C_i = \begin{bmatrix} 1\,0\,0\,0 \\ 0\,0\,0\,0 \end{bmatrix}, \text{ Sensor A}$$
$$C_i = \begin{bmatrix} 0\,0\,0\,0 \\ 0\,0\,1\,0 \end{bmatrix}, \text{ Sensor B}$$
$$C_i = \begin{bmatrix} 0\,0\,0\,0 \\ 0\,0\,0\,0 \end{bmatrix}, \text{ Non-Sensor Node.}$$

Obviously, the sensor network is collectively observable. In the first experiment, we take the number of consensus steps $L = 3$. To check whether the stochastic boundedness in Theorem 10.3.2 is satisfied, we run 1000 independent simulations and then count the occurrence of the event $\sum_{i=1}^{N} \|P_k^i\| > \epsilon$ with $\epsilon = 300N$. The result in Table 10.2 shows that $\mathbb{P}(\sum_{i=1}^{N} \|P_k^i\| > \epsilon) = 0$ and hence the covariance is boundedness. The behaviors of the average root-mean-square errors (ARMSEs) of the Kalman-consensus filter are shown in Fig. 10.2. As it can be seen, the filtering performance depends crucially on the link failure rate. Moreover, we compare the ARMSEs of the Kalman-consensus filter under different measurement noise variances in Fig. 10.3. It can be seen that a higher noise variance will decrease the filtering performance. In the second experiment, we let $\mathbb{P}(\lambda_{i,j}^{k,l} = 1) = 0.9$ and choose a sufficiently large number of consensus steps ($L = 500$). The simulation result is plotted in Fig. 10.4 which confirms that the covariance is indeed bounded.

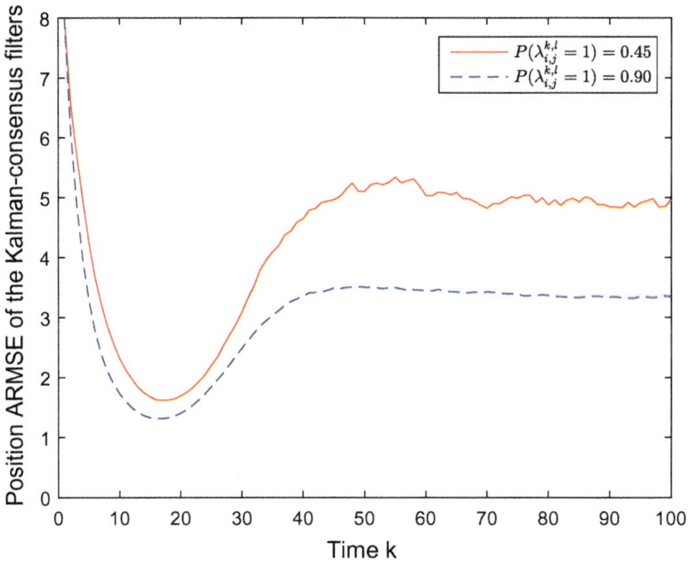

Fig. 10.2 Position ARMSE of the Kalman-consensus filters under difference link failure rates. The number of consensus steps is chosen as $L = 1$

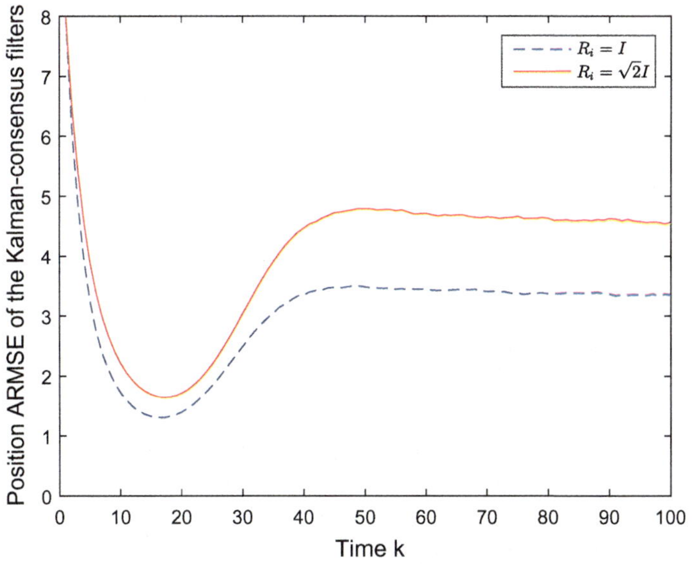

Fig. 10.3 Position ARMSE of the Kalman-consensus filters under difference measurement noise variances. The number of consensus steps is chosen as $L = 1$

Table 10.2 The incidence rate $\mathbb{P}\left(\sum_{i=1}^{N} \|P_k^i\| > \epsilon\right)$

ϵ	500	600	700	1000
Percent (%)	48.3	1.9	0.1	0

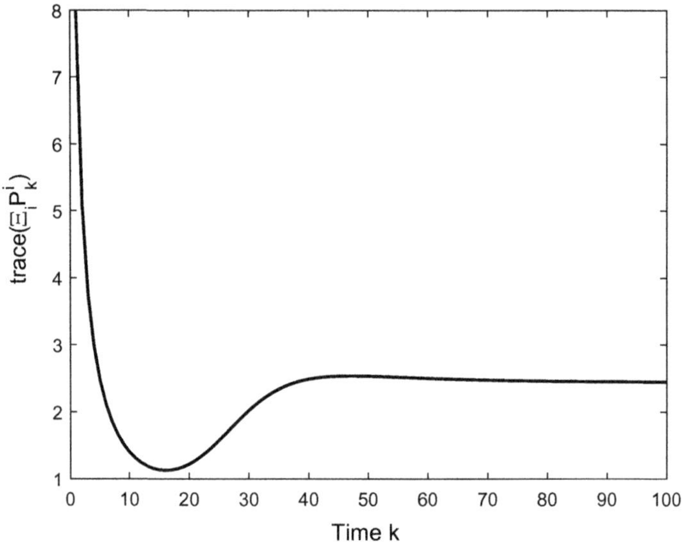

Fig. 10.4 Variance of the position error tr $\left(\Xi P_k^i\right)$, $i = 1, \ldots, 70$, where $\Xi = \text{diag}\{1, 0, 1, 0\}$

10.5 Conclusions

In this chapter, we have investigated the distributed state estimation problems under random communication link failures. A two-stage Kalman-consensus filtering approach consisting of a standard Kalman filter update and a consensus update has been introduced. The consensus procedure adopts the covariance intersection technique with the hope to improve the filtering performance. The analysis of stochastic boundedness has been carried out, respectively, in finite and infinite consensus step situations. It has been proved that, when the collective observability and network connectivity satisfy certain conditions, the error covariance of the Kalman-consensus filter is stochastically bounded in distribution. Furthermore, if the number of consensus steps is chosen to be infinite, it has been shown that the uniform boundedness is achieved almost surely. At last, the validity of the proposed Kalman-consensus filter has been assessed through a numerical example.

References

1. I. Akyildiz, W. Su, Y. Sankarasubramaniam, E. Cayirci, Wireless sensor networks: a survey. Comput. Netw. **38**(4), 393–422 (2002)
2. L. Wang, Z. Wang, T. Huang, G. Wei, An event-triggered approach to state estimation for a class of complex networks with mixed time delays and nonlinearities. IEEE Trans. Cybern. **46**(11), 2497–2508 (2016)
3. J. Li, H. Dong, F. Han, N. Hou, X. Li, Filter design, fault estimation and reliable control for networked time-varying systems: a survey. Syst. Sci. Control Eng. Open Access J. **5**(1), 331–341 (2017)
4. K. You, L. Xie, S. Song, Asymptotically optimal parameter estimation with scheduled measurements. IEEE Trans. Signal Process. **61**(14), 3521–3531 (2013)
5. Y. Luo, Z. Wang, G. Wei, F.E. Alsaadi, Robust H_∞ filtering for a class of two-dimensional uncertain fuzzy systems with randomly occurring mixed delays. IEEE Trans. Fuzzy Syst. **25**(1), 70–83 (2017)
6. D. Ding, Z. Wang, G. Wei, F.E. Alsaadi, Event-based security control for discrete-time stochastic systems. IET Control Theory Appl. **10**(15), 1808–1815 (2016)
7. Q. Liu, Z. Wang, X. He, D.H. Zhou, Event-based H_∞ consensus control of multi-agent systems with relative output feedback: the finite-horizon case. IEEE Trans. Autom. Control **60**(9), 2553–2558 (2015)
8. Z. Deng, Y. Gao, L. Mao, Y. Li, G. Hao, New approach to information fusion steady-state Kalman filtering. Automatica **41**(10), 1695–1707 (2005)
9. D. Willner, C. Chang, K. Dunn, Kalman filter algorithms for a multi-sensor system, in *Proceedings of 15th IEEE Conference on Decision and Control*, Clearwater, Florida (1976), pp. 570–574
10. I. Matei, J. Baras, Consensus-based linear distributed filtering. Automatica **48**(8), 1776–1782 (2012)
11. R. Olfati-Saber, Distributed Kalman filtering for sensor networks, in *Proceedings of 46th IEEE Conference on Decision and Control*, New Orleans, LA, USA (2007), pp. 5492–5498
12. S. Zhu, Y. Soh, L. Xie, Distributed inference for relay-assisted sensor networks with intermittent measurements over fading channels. IEEE Trans. Signal Process. **64**(3), 742–756 (2016)
13. A. Simonetto, T. Keviczky, R. Babuška, Distributed nonlinear estimation for robot localization using weighted consensus, in *2010 IEEE International Conference on Robotics and Automation*, Alaska, USA (2010), pp. 3026–3031
14. L. Ma, Z. Wang, Q.-L. Han, H.K. Lam, Variance-constrained distributed filtering for time-varying systems with multiplicative noises and deception attacks over sensor networks. IEEE Sens. J. **17**(7), 2279–2288 (2017)
15. Q. Liu, Z. Wang, X. He, G. Ghinea, F.E. Alsaadi, A resilient approach to distributed filter design for time-varying systems under stochastic nonlinearities and sensor degradation. IEEE Trans. Signal Process. **65**(5), 1300–1309 (2017)
16. Q. Liu, Z. Wang, X. He, D.H. Zhou, Event-based recursive distributed filtering over wireless sensor networks. IEEE Trans. Autom. Control (2015)
17. D. Ding, Z. Wang, D.W.C. Ho, G. Wei, Distributed recursive filtering for stochastic systems under uniform quantizations and deception attacks through sensor networks. Automatica **78**, 231–240 (2017)
18. S. Das, J. Moura, Distributed Kalman filtering with dynamic observations consensus. IEEE Trans. Signal Process. **63**(17), 4458–4473 (2015)
19. G. Battistelli, L. Chisci, C. Fantacci, A. Farina, A. Graziano, Consensus CPHD filter for distributed multitarget tracking. IEEE J. Sel. Topics Signal Process. **7**(3), 508–520 (2013)
20. L. Paradis, Q. Han, A survey of fault management inwireless sensor networks. J. Netw. Syst. Mgt. **5**(2), 171–190 (2007)
21. Y. Yuan, H. Yuan, L. Guo, H. Yang, S. Sun, Resilient control of networked control system under DoS attacks: a unified game approach. IEEE Trans. Ind. Inform. **12**(5), 1786–1794 (2016)
22. Y. Hatano, M. Mesbahi, Agreement over random networks. IEEE Trans. Autom. Control **50**(11), 1867–1872 (2005)

23. A. Tahbaz-Salehi, A. Jadbabaie, Consensus over ergodic stationary graph processes. IEEE Trans. Autom. Control **55**(1), 225–230 (2010)
24. G. Battistelli, L. Chisci, G. Mugnai, A. Farina, A. Graziano, Consensus-based linear and nonlinear filtering. IEEE Trans. Autom. Control **60**(5), 1410–1415 (2015)
25. G. Calafiore, F. Abrate, Distributed linear estimation over sensor networks. Int. J. Control **82**(5), 868–882 (2009)
26. L. Xiao, S. Boyd, Fast linear iterations for distributed averaging. Syst. Control Lett. **53**(1), 65–78 (2004)
27. W. Niehsen, Information fusion based on fast covariance intersection filtering, in *IEEE Proceedings of the 5th International Conference on Information Fusion*, Maryland, vol. 2 (2002), pp. 901–904
28. B. Sinopoli, L. Schenato, M. Franceschetti, K. Poolla, M. Jordan, S. Sastry, Kalman filtering with intermittent observations. IEEE Trans. Autom. Control **49**(9), 1453–1464 (2004)
29. D. Alonso-Román, B. Beferull-Lozano, Adaptive consensus-based distributed Kalman filter for WSNs with random link failures, in *IEEE Conference on Distributed Computing in Sensor Systems* (2016), pp. 187–192
30. S. Das, J. Moura, Consensus+innovations distributed kalman filter with optimized gains. IEEE Trans. Signal Process. **65**(2), 467–481 (2017)

Chapter 11
Moving-Horizon Estimation with Binary Encoding Schemes

Complex networks are composed of a large number of dynamical nodes interconnected according to network topologies. Many complicated practical systems can be generally described by complex networks in terms of nodes, edges, and interactions. Thanks to their extensive applications in diverse real-world systems such as electrical power systems, manufacturing processes, compartmental systems, and biological processes, the analysis and synthesis problems of complex networks have now become a very active research topic in both industry and academia.

In recent years, considerable research efforts have been devoted to stabilization and synchronization problems of complex networks, see, for example, [1]. Note that the conventional closed-loop controller design approach usually adopts the full state feedback techniques, thereby requiring knowledge of the full state. However, owing to the technological restrictions of the sensing devices, the full state might be unavailable in many real-world systems. A proper solution is to make use of state estimators to extract the system state from the partial observations. For this reason, the state estimation problems of complex networks have recently become a research focus and various results have been available in the literature, see, e.g., [2, 3]. To be specific, the recursive state estimation problems for time-varying complex networks have been investigated in [2], where an extended-Kalman-filter-like technique is exploited to generate a performance-guaranteed estimate. Moreover, to deal with the shortcomings of the conventional centralized filtering algorithms such as heavy computational requirements, the authors in [3] have developed the distributed Kalman filter for complex large-scale dynamical systems.

Since the pioneering work [4], the Kalman filter has served as one of the most successful technologies in signal processing field. This is because it provides the optimal estimate in the minimum mean-square error sense for linear Gauss–Markov systems. We however remark that there are some limitations of the Kalman filter. For example, the Gaussian noise assumption is necessary for the optimality of the Kalman filter, while it mismatches with many practical applications wherein the noise distributions are non-Gaussian or even unknown. In addition, uncertainties in modeling processes also prevent one from successfully applying the Kalman filter.

Q. Liu et al., *Stochastic Control and Filtering over Constrained Communication Networks*, Studies in Systems, Decision and Control 178,
https://doi.org/10.1007/978-3-030-00157-5_11

To overcome the aforementioned limitations, a moving-horizon estimation strategy has been proposed recently. The underlying methodologies are to reconstruct the system states according to a sequence of past measurements in a moving but fixed-size window. For instance, [5, 6] have investigated the moving-horizon estimation problems for linear discrete-time systems with disturbance constraints. To deal with the measurement outliers, a robust moving-horizon strategy has been established in [7] via solving a set of least-square optimization problems. The related results on linear systems have been further extended to the nonlinear systems in [8, 9].

With the rapid development of communication technologies, the remote state estimation in networked systems has become a cutting-edge research. Although the use of network equipments provides flexible architecture, reduced installation, and maintenance costs, it also poses serious challenges on the state estimation problems. For example, due to the inherent limitations of digital communication channels, the measurements transmitted from the sensor to the estimator can be affected by certain network-induced phenomena, which might result in a dramatic performance degradation. As such, many attempts have been concentrated on addressing the estimation problems under various communication situations over the last few years, see, e.g., the quantization effects in [10], and the multiple packet dropouts in [11]. On the other hand, as noted by [12], the binary encoding is one of the most widely employed schemes in the digital communication processes for the reason that the binary data is robust in transmission and simple to implement. Under the binary encoding scheme, signals shall be first encoded into a group of binary bits and then forwarded via binary symmetric channels (BSCs), where the binary bits might suffer from random bit errors induced by the channel noises. Some initial studies have been devoted to investigating the binary encoding scheme from the perspective of information theory, see, e.g., [13]. Unfortunately, when it comes to state estimation problems, some important issues still remain obscure. For instance, how to design the remote estimator that adopts encoded signals with bit errors and what is the influence of the encoding scheme on the estimation performance. Therefore, it is of practical importance to carry out an initial study on moving-horizon estimation problems under binary encoding schemes.

Motivated by the above discussion, in this chapter, we aim to develop moving-horizon estimation algorithms for discrete-time complex networks under the binary encoding schemes. First, centralized moving-horizon estimation problems are investigated, where all the nodes forward their local measurements to a central unit in order to obtain a global optimal estimate. Second, decentralized moving-horizon estimation problems are further studied, where the individual node exploits the measurement from itself and the estimates received from its neighbors to give a local optimal estimate. The main contributions of this chapter can be highlighted as follows: (1) The signal distortions in communication processes under the binary encoding scheme are quantified; (2) novel centralized/decentralized moving-horizon estimators are established respectively with specific efforts to compensate the signal distortions, and (3) sufficient conditions are obtained for the stochastically ultimate boundedness of the estimation error.

The remainder of the chapter is structured as follows. In Sect. 11.1, the complex networks and the binary encoding scheme are introduced. In Sect. 11.2, some preliminary knowledge of the binary encoding scheme is presented. The centralized and decentralized moving-horizon estimators are constructed in Sects. 11.3 and 11.4, respectively. Section 11.5 shows the simulation results and the associated discussion. The conclusions are drawn in Sect. 11.6.

11.1 System Description

11.1.1 Complex Networks

The interconnections in complex networks can be described by a directed graph, denoted by $\mathcal{G} \triangleq (\mathcal{V}, \mathcal{E}, B)$, with the set of nodes $\mathcal{V} = \{\mathcal{S}_1, \mathcal{S}_2, \ldots, \mathcal{S}_n\}$, the set of edges $\mathcal{E} = \mathcal{V} \times \mathcal{V}$, and the weighted adjacency matrix $B = [b_{ij}]_{n\times n}$. The elements b_{ij} in the weighted adjacency matrix are nonnegative and satisfy the property $b_{ij} > 0 \Longleftrightarrow (\mathcal{S}_i, \mathcal{S}_j) \in \mathcal{E}$, which means the dynamics of ith node can be affected by jth node if and only if $b_{ij} > 0$. We assume that the self-loop does not exist throughout this chapter, that is, $b_{ii} = 0$, for $i = 1, 2, \ldots, n$. The neighbors of ith node are denoted by $\mathcal{N}_i \triangleq \{j|(\mathcal{S}_i, \mathcal{S}_j) \in \mathcal{E}\}$.

Consider a discrete-time complex network consisting of n linearly coupled heterogeneous nodes as follows:

$$\begin{cases} x_{i,k+1} = A_i x_{i,k} + \sum\limits_{j\in\mathcal{N}_i} b_{ij}\Gamma x_{j,k} + w_{i,k} \\ y_{i,k} = C_i x_{i,k} + v_{i,k}, \ \text{for } i = 1, 2, \ldots, n, \end{cases} \tag{11.1}$$

where $x_{i,k} \in \mathbb{R}^{n_x}$ and $y_{i,k} \in \mathbb{R}^{n_y}$ are the state vector and the measurement output of node i. $w_{i,k} \in W \subset \mathbb{R}^{n_x}$ and $v_{i,k} \in V \subset \mathbb{R}^{n_y}$ are the system noise and the measurement noise of node i. W and V are polyhedral and convex sets with $0 \in W$ and $0 \in V$. The inner coupling strength $\Gamma \in \mathbb{R}^{n_x\times n_x}$ is a positive definite diagonal matrix. The transition matrix A_i and measurement matrix C_i are known matrices of appropriate dimensions.

11.1.2 Binary Encoding Schemes

As shown in Fig. 11.1, the binary encoding schemes are taken into consideration during signal transmission. In this case, the original signals are encoded into a finite-length binary bit string, and then transmitted to a remote estimator for further processing via a memoryless BSC.

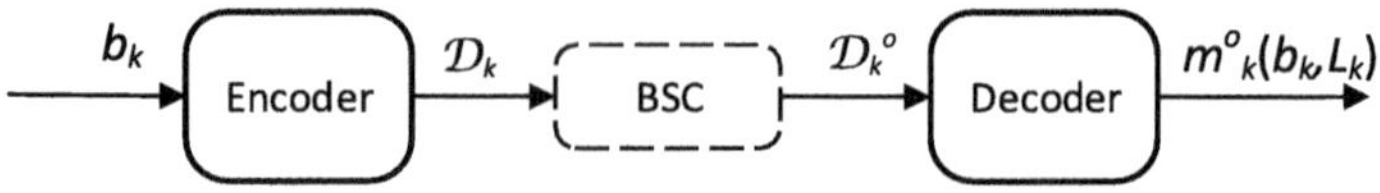

Fig. 11.1 The binary encoding schemes

Suppose that the range of the scalar signal $b_k \in \mathbb{R}$ at instant k is $[-\hbar, \hbar]$, where $\hbar \in \mathbb{R}$ is an application dependent positive scalar. A binary encoder is utilized to convert the signal b_k into a binary bit string of length L. We therefore have 2^L points denoted by

$$\mathcal{U} \triangleq \{\tau_1, \tau_2, \ldots, \tau_{2^L}\}.$$

These points are uniformly spaced, which divide the whole range into $2^L - 1$ segments with a uniform interval length $\delta = \tau_{i+1} - \tau_i$, for $i = 1, 2, \ldots, 2^L - 1$. Moreover, it can be seen that

$$\delta = \frac{2\hbar}{2^L - 1}. \tag{11.2}$$

First, a stochastic truncation function is utilized to pretreat the signal b_k as follows:

$$\mathcal{Q}_k : b_k \to m_k(b_k, L),$$

where $m_k(b_k, L)$ is the truncated output. When $\tau_i \leq b_k \leq \tau_{i+1}$, the output $m_k(b_k, L)$ is generated according to the following probabilistic manner:

$$\begin{cases} \mathbb{P}\{m_k(b_k, L) = \tau_i\} = 1 - r_k \\ \mathbb{P}\{m_k(b_k, L) = \tau_{i+1}\} = r_k, \end{cases} \tag{11.3}$$

where $r_k \triangleq (b_k - \tau_i)/\delta$ and $0 \leq r_k \leq 1$. Furthermore, the output $m_k(b_k, L)$ can be represented on a basis of binary bits as follows:

$$m_k(b_k, L) = -\hbar + \sum_{i=1}^{L} h_{i,k} 2^{i-1} \delta. \tag{11.4}$$

Now, we have encoded b_k into the following binary bit string:

$$\mathcal{D}_k \triangleq \{h_{1,k}, h_{2,k}, \ldots, h_{L,k}\}, \ h_{i,k} \in \{0, 1\}.$$

The next step is to transmit the binary bit strings $\mathcal{D}_k$ by a memoryless BSC, where every bit might flip with a small probability (called crossover probability hereafter) because of channel noises. Accordingly, the received bit string is defined by

$$\mathcal{D}_k^o \triangleq \{h_{1,k}^o, h_{2,k}^o, \ldots, h_{L,k}^o\},\ h_{i,k}^o \in \{0, 1\},$$

where $h_{i,k}^o = \theta_{i,k}(1 - h_{i,k}) + (1 - \theta_{i,k})h_{i,k}$ with

$$\theta_{i,k} = \begin{cases} 1, & \text{the } i\text{th bit is flipped} \\ 0, & \text{the } i\text{th bit is not flipped.} \end{cases} \tag{11.5}$$

Let the crossover probability be p, and then we have

$$\mathbb{P}\{\theta_{i,k} = 1\} = p.$$

For the convenience of analysis, we assume that $\theta_{i,k}$ are white and mutually independent in this chapter. Finally, the received bit string $\mathcal{D}_k^o$ can be decoded to restore the original signals according to the following equation:

$$m_k^o(b_k, L) = -\hbar + \sum_{i=1}^{L} h_{i,k}^o 2^{i-1}\delta. \tag{11.6}$$

Owing to constrained communication bandwidth in practice, only a finite bit budget can be utilized to encode the signal over communication channels, which makes it necessary to pretreat the signals by using the truncation function. Generally speaking, under the presented binary encoding scheme, the transmission error mainly stems from two aspects: truncation errors and flipped bit errors. To illustrate the binary encoding scheme, an example is presented as follows.

Example 1 Suppose that the original signal $b_k = -0.40$. Choose the range $\hbar = 0.75$ and the length of the string $L = 4$. In view of (11.2), the interval length can be obtained as $\delta = 0.1$. By utilizing the truncation function (11.3), we have the output $m_k(b_k, L) = -0.35$ or -0.45 with equal probability, i.e., $r_k = 0.5$. Assume that, in this sample path, $m_k(b_k, L) = -0.35$, which therefore can be encoded into the bit string $\mathcal{D}_k = \{0010\}$ and then forwarded via BSCs. When the first bit of the string suffers from the bit error, we will get $\mathcal{D}_k^o = \{1010\}$ at the receiver. According to the decoder given in (11.6), we can recover the signal as $m_k^o(b_k, L) = -0.25$.

11.2 Preliminary

In this section, some preliminary knowledge of the truncation function as well as the memoryless BSC are introduced.

The output can be rewritten by

$$m_k(b_k, L) = b_k + q_k \tag{11.7}$$

where $q_k \triangleq m_k(b_k, L) - b_k$ represents the truncation error. According to (11.3), we know that q_k is a stochastic noise obeying the Bernoulli distribution taking values at $r_k\delta$ or $(r_k - 1)\delta$, i.e.,

$$\begin{cases} \mathbb{P}\{q_k = -r_k\delta\} = 1 - r_k \\ \mathbb{P}\{q_k = (1 - r_k)\delta\} = r_k \end{cases} \tag{11.8}$$

The following lemma presents some statistical properties of the truncation error.

Lemma 11.2.1 ([12]) *The truncation error q_k is with zero mean and bounded variance, that is,*

$$\mathbb{E}\{q_k\} = 0, \quad \mathbb{E}\{q_k^2\} \le \frac{\delta^2}{4}. \tag{11.9}$$

An important property of BSCs is presented as follows.

Lemma 11.2.2 *Assume that the signal $m_k(b_k, L)$ is transmitted via a memoryless BSC with crossover probability p. Then, the received signal $m_k^o(b_k, L)$ is with the mean and variance given by*

$$\mathbb{E}\big\{m_k^o(b_k, L)\big\} = (1 - 2p)m_k(b_k, L)$$

and

$$\mathrm{Var}\big\{m_k^o(b_k, L)\big\} = p(1 - p)\frac{4\hbar^2(2^{2L} - 1)}{3(2^L - 1)^2},$$

where the expectation is taken with respect to the random variables $\theta_{i,k}$.

Proof Taking mathematical expectation of $m_k^o(b_k, L)$ over the random variables $\theta_{i,k}$ yields that

$$\begin{aligned}
&\mathbb{E}\big\{m_k^o(b_k, L)\big\} \\
&= -\hbar + \sum_{i=1}^{L} \mathbb{E}\{h_{i,k}^o\}2^{i-1}\delta \\
&= -\hbar + \sum_{i=1}^{L} \Big(p(1 - h_{i,k}) + (1 - p)h_{i,k}\Big)2^{i-1}\delta \\
&= m_k(b_k, L) + p\sum_{i=1}^{L} \Big(1 - 2h_{i,k}\Big)2^{i-1}\delta,
\end{aligned}$$

where the last equality follows from $m_k(b_k, L) = -\hbar + \sum_{i=1}^{L} h_{i,k}2^{i-1}\delta$. In addition, we have

$$
\begin{aligned}
&p\sum_{i=1}^{L}\Big(1-2h_{i,k}\Big)2^{i-1}\delta\\
&=p\Big(\sum_{i=1}^{L}2^{i-1}\delta-2\sum_{i=1}^{L}h_{i,k}2^{i-1}\delta\Big)\\
&=2p\Big(\hbar-\sum_{i=1}^{L}h_{i,k}2^{i-1}\delta\Big)\\
&=-2pm_k(b_k,L).
\end{aligned}
$$

It is therefore straightforward to verify that

$$
\mathbb{E}\big\{m_k^o(b_k,L)\big\}=(1-2p)m_k(b_k,L).
$$

Moreover, the variance of $m_k^o(b_k, L)$ can be derived as follows:

$$
\begin{aligned}
&\mathrm{Var}\big\{m_k^o(b_k,L)\big\}\\
&=\mathbb{E}\left\{\Big(-\hbar+\sum_{i=1}^{L}h_{i,k}^o2^{i-1}\delta\Big)^2\right\}-\Big(\mathbb{E}\big\{m_k^o(b_k,L)\big\}\Big)^2\\
&=\mathbb{E}\left\{\Big(\sum_{i=1}^{L}(h_{i,k}^o-\mathbb{E}\{h_{i,k}^o\})2^{i-1}\delta+\mathbb{E}\big\{m_k^o(b_k,L)\big\}\Big)^2\right\}\\
&\quad-\Big(\mathbb{E}\big\{m_k^o(b_k,L)\big\}\Big)^2\\
&=\mathbb{E}\left\{\Big(\sum_{i=1}^{L}(h_{i,k}^o-\mathbb{E}\{h_{i,k}^o\})2^{i-1}\delta\Big)^2\right\}.
\end{aligned}
$$

Noting that $\theta_{i,k}$, for $i = 1, 2, \ldots, n$, are mutually independent, we can see $h_{i,k}^o$ are also mutually independent, which, together with the fact that $h_{i,k} \in \{0, 1\}$, further indicates that

$$
\begin{aligned}
&\mathbb{E}\left\{\Big(\sum_{i=1}^{L}(h_{i,k}^o-\mathbb{E}\{h_{i,k}^o\})2^{i-1}\delta\Big)^2\right\}\\
&=\sum_{i=1}^{L}\Big(\mathbb{E}\left\{(h_{i,k}^o)^2\right\}-(\mathbb{E}\{h_{i,k}^o\})^2\Big)2^{2i-2}\delta^2\\
&=p(1-p)\frac{4\hbar^2(2^{2L}-1)}{3(2^L-1)^2}.
\end{aligned}
$$

The proof is complete now.

Although the binary encoding scheme has been established for scalar signals so far, we should emphasize that it is also applicable to the vector signals in a component-wise manner, i.e., $m_k(b_k, L) = \text{vec}\{m_k(b_{1,k}, L), \ldots, m_k(b_{n,k}, L)\}$ with $b_{i,k}$ being the ith entry of the vector $b_k \in \mathbb{R}^n$. For compactness, the aforementioned notations that define the scalar signals can be extended to represent the vectors. For example, when $b_k \in \mathbb{R}^n$, we have $m_k(b_k, L) \in \mathbb{R}^n$, $m_k^o(b_k, L) \in \mathbb{R}^n$, and $q_k \in \mathbb{R}^n$. Moreover, since the stochastic signal truncation is carried out independently for every component of b_k, the entries in q_k are mutually independent.

11.3 Centralized Moving-Horizon Estimation

In this section, a centralized moving-horizon estimation approach is taken into consideration, as shown in Fig. 11.2, where the central unit has access to the information of all the nodes. To be more specific, we assume that the local measurements $y_{i,k}$ are forwarded to the center unit using the binary encoding scheme, and then the center gives an estimate of all the plant states at each time instant.

According to Lemmas 11.2.1 and 11.2.2, it can be seen that, compared with the original signals $y_{i,k}$, the received ones $m_k^o(y_{i,k}, L)$ are inevitably subject to certain degree of distortions. In order to compensate for the distortions, we adopt the recovered measurements as follows:

$$\tilde{y}_{i,k} = \gamma_k^{-1} m_k^o(y_{i,k}, L), \tag{11.10}$$

where $\gamma_k \triangleq 1 - 2p$, such that the means of the recovered signals are equal to the original ones, i.e., $\mathbb{E}\{\tilde{y}_{i,k}\} = y_{i,k}$. As a consequence, the equivalent noise stemming from the bit error in BSCs can be denoted by

$$\tilde{m}_{i,k} \triangleq \tilde{y}_{i,k} - m_k(y_{i,k}, L). \tag{11.11}$$

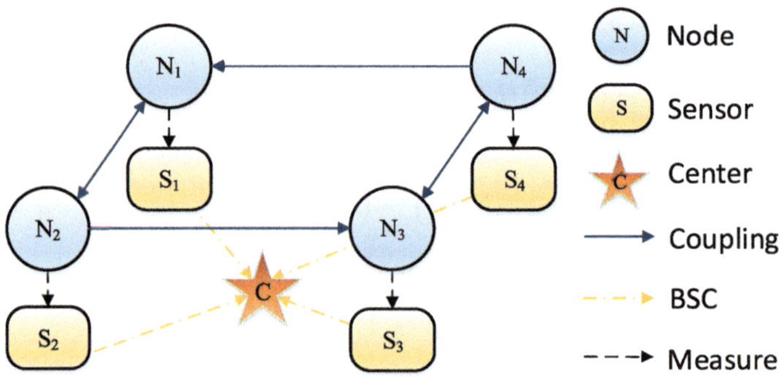

Fig. 11.2 Centralized moving-horizon estimation

Similarly to (11.7), we denote the truncation error $q_{i,k} \triangleq m_k(y_{i,k}, L) - y_{i,k}$, and thereby have

$$\tilde{y}_{i,k} \triangleq y_{i,k} + \tilde{m}_{i,k} + q_{i,k}. \tag{11.12}$$

In what follows, we will consider the moving-horizon estimation for complex networks with binary encoding schemes. The fundamental methodology of moving-horizon estimation is to solve an optimization problem over a moving but fixed-size horizon. More specifically, at each stage $k = N, N+1, \ldots$, our purpose is to generate an estimate of all the target plants $x_{i,k-N}, \ldots, x_{i,k}$ $(i = 1, 2, \ldots, n)$, denoted by $\hat{x}_{i,k-N|k}, \ldots, \hat{x}_{i,k|k}$, based on the history measurements $\tilde{y}_{i,k-N}, \ldots, \tilde{y}_{i,k}$ as well as the prediction $\bar{x}_{i,k-N}$ of the state $x_{i,k-N}$ $(i = 1, 2, \ldots, n)$. A natural way of developing the prediction is to make use of the dynamical equation (11.1) while replacing the unknown states by the corresponding real-time estimates, i.e.,

$$\bar{x}_{i,k-N} = A_i \hat{x}_{i,k-N-1|k-1} + \sum_{j \in \mathcal{N}_i} b_{ij} \Gamma \hat{x}_{j,k-N-1|k-1}. \tag{11.13}$$

To make the formulas more compact, we introduce new notations to stack a set of vectors $\upsilon_{i,k}$ $(i = 1, 2, \ldots, n)$ in the following analysis, that is, $\upsilon_k \triangleq \mathrm{vec}\{\upsilon_{1,k}, \upsilon_{2,k}, \ldots, \upsilon_{n,k}\}$ and $\upsilon_{k-N}^k \triangleq \mathrm{vec}\{\upsilon_{k-N}, \upsilon_{k-N+1}, \ldots, \upsilon_k\}$. Here, υ is generic that could be x, y, $\tilde{y}$, q, $\tilde{m}$, v, w, $\bar{x}$, or $\hat{x}$.

Now, we can state the centralized moving-horizon estimation problem as follows.

Problem 1 At each time instant k, for given information $(\hat{x}_{k-N|k-1}, \tilde{y}_{k-N}^k)$, drive the optimal state estimate $\hat{x}_{k-N|k}, \ldots, \hat{x}_{k|k}$ by minimizing a least-square cost function defined as follows:

$$\begin{aligned} J_k = \epsilon \sum_{i=1}^{n} \|\hat{x}_{i,k-N|k} - \bar{x}_{i,k-N}\|^2 \\ + \sum_{t=k-N}^{k} \sum_{i=1}^{n} \|\tilde{y}_{i,t} - \hat{y}_{i,t|k}\|^2 \end{aligned} \tag{11.14}$$

subject to the following constraints:

$$\begin{cases} \hat{x}_{i,t+1|k} = A_i \hat{x}_{i,t|k} + \sum_{j=1}^{n} b_{ij} \Gamma \hat{x}_{j,t|k} \\ \hat{y}_{i,t|k} = C_i \hat{x}_{i,t|k}, \end{cases} \tag{11.15}$$

for $t = k - N, \ldots, k$, where ϵ is a positive scalar.

Remark 11.1 In Problem 1, the term $\|\hat{x}_{i,k-N|k} - \bar{x}_{i,k-N}\|^2$ is the initial penalty representing our beliefs in the one-step prediction, and the term $\|\tilde{y}_{i,t} - \hat{y}_{i,t|k}\|^2$ penalties the difference between the estimates and the measurements. As the one-step prediction is made on a basis of the history data, the parameter ϵ in fact is utilized to weigh the influence of the most recent N step measured data and that of the history data.

Remark 11.2 As stated previously, the moving-horizon estimation problems have been widely studied for discrete-time linear systems. However, we should emphasize that the classical moving-horizon estimation strategies are no longer applicable to this chapter as signals suffer from signal truncation errors and stochastic bit errors induced by communication processes. These errors might dramatically degrade the estimation accuracy, and thereby call for urgent efforts to develop a novel moving-horizon estimation strategy for practical applications.

Before proceeding further, let us denote

$$F_N \triangleq \begin{bmatrix} C \\ C\bar{A} \\ \vdots \\ C\bar{A}^N \end{bmatrix}, \ H_N \triangleq \begin{bmatrix} 0 & 0 & \cdots & 0 & 0 \\ C & 0 & \cdots & 0 & 0 \\ C\bar{A} & C & \cdots & 0 & 0 \\ \vdots & \vdots & \cdots & \vdots & \vdots \\ C\bar{A}^{N-1} & C\bar{A}^{N-2} & \cdots & C\bar{A} & C \end{bmatrix}$$

$$A \triangleq \text{diag}\{A_1, A_2, \dots, A_n\}, \quad C \triangleq \text{diag}\{C_1, C_2, \dots, C_n\},$$

$$\bar{A} \triangleq A + B \otimes \Gamma, \quad \Delta \triangleq \frac{2\hbar\sqrt{p(1-p)(2^{2L}-1)}}{(1-2p)(2^L-1)\sqrt{3}}.$$

Utilizing these notations, we are in a position to present the following theorem.

Theorem 11.3.1 *For given information $\left(\hat{x}_{k-N|k-1}, \tilde{y}_{k-N}^k\right)$, Problem 1 has a unique optimal solution as follows:*

$$\hat{x}_{k-N|k} = (\epsilon I + F_N' F_N)^{-1}(\epsilon \bar{x}_{k-N} + F_N' \tilde{y}_{k-N}^k). \tag{11.16}$$

Moreover, the corresponding estimation error $e_{k-N} \triangleq x_{k-N} - \hat{x}_{k-N|k}$ is given by

$$\begin{aligned} e_{k-N} = {} & (\epsilon I + F_N' F_N)^{-1} \\ & \times \Big(\epsilon \bar{A} e_{k-N-1} + \epsilon w_{k-N-1} - F_N' H_N w_{k-N}^{k-1} \\ & - F_N' v_{k-N}^k - F_N' \tilde{m}_{k-N}^k - F_N' q_{k-N}^k\Big). \end{aligned} \tag{11.17}$$

Proof The necessary condition on the minimum of the cost function (11.14) is

$$\begin{aligned} \nabla_{\hat{x}_{k-N|k}} J(k) = {} & 2\epsilon(\hat{x}_{k-N|k} - \bar{x}_{k-N}) \\ & - 2F_N'(\tilde{y}_{k-N}^k - F_N \hat{x}_{k-N|k}) = 0. \end{aligned}$$

Moreover, it is easy to show that the Hessian matrix of the cost function is positive definite, and hence the cost function can be minimized by choosing the estimate as follows:

$$\hat{x}_{k-N|k} = (\epsilon I + F_N' F_N)^{-1}(\epsilon \bar{x}_{k-N} + F_N' \tilde{y}_{k-N}^k).$$

Furthermore, subtracting (11.16) from (11.1), one has

$$e_{k-N} = x_{k-N} - (\epsilon I + F_N' F_N)^{-1}(\epsilon \bar{x}_{k-N} + F_N' \tilde{y}_{k-N}^k). \tag{11.18}$$

Based on (11.11), the measurements can be rewritten as follows:

$$\tilde{y}_{k-N}^k = \tilde{m}_{k-N}^k + y_{k-N}^k + q_{k-N}^k. \tag{11.19}$$

Substituting the one-step prediction (11.13) and the measurements (11.19) into (11.18) yields that

$$\begin{aligned} e_{k-N} = & (\epsilon I + F_N' F_N)^{-1} \\ & \times \Big(\epsilon \bar{A} e_{k-N-1} + \epsilon w_{k-N-1} + F_N' F_N x_{k-N} \\ & - F_N' \big(\tilde{m}_{k-N}^k + y_{k-N}^k + q_{k-N}^k\big)\Big). \end{aligned}$$

Noting that

$$y_{k-N}^k = F_N x_{k-N} + H_N w_{k-N}^{k-1} + v_{k-N}^k,$$

we can get the dynamics of the estimation error as follows:

$$\begin{aligned} e_{k-N} = & (\epsilon I + F_N' F_N)^{-1} \\ & \times \Big(\epsilon \bar{A} e_{k-N-1} + \epsilon w_{k-N-1} - F_N' H_N w_{k-N}^{k-1} \\ & - F_N' v_{k-N}^k - F_N' \tilde{m}_{k-N}^k - F_N' q_{k-N}^k\Big), \end{aligned}$$

which completes the proof.

Different from the standard moving-horizon estimation proposed in [5], the stochastic behaviors of the truncation error and the flipped bit error make the estimation error (11.17) to be a stochastic sequence. As a consequence, we need to investigate the statistical properties of the error sequence $\{e_k\}$ in what follows. For this purpose, the following definition is introduced.

Definition 11.3.2 The error dynamics governed by (11.17) is said to be *stochastically ultimately bounded* if there exist scalars $0 < \mu_1 < 1$ and $\mu_2 > 0$ such that

$$\mathbb{E}\big\{\|e_{k-N}\|\big\} \leq \mu_1^{k-N} \mathbb{E}\big\{\|e_0\|\big\} + \mu_2.$$

Moreover, for the sake of clarity, we denote

$$\bar{a} \triangleq \|\bar{A}\|, \quad \bar{f} \triangleq \|F_N\|, \quad \bar{f}_{\min} \triangleq \|F_N\|_{\min},$$
$$\bar{w} \triangleq \max_{w_{i,k}\in W} \|w_{i,k}\|, \quad \bar{v} \triangleq \max_{v_{i,k}\in V} \|v_{i,k}\|.$$

Now, we are ready to state the following results.

Theorem 11.3.3 *The dynamics of the estimation error* (11.17) *is stochastically ultimately bounded if* $0 < c_1 < 1$. *Moreover, the upper bound of the expectation of the estimate error can be given by*

$$\lim_{k\to\infty} \mathbb{E}\{\|e_k\|\} \leq \frac{c_\infty}{\epsilon(1-\bar{a}) + \bar{f}_{\min}^2}, \tag{11.20}$$

where

$$\begin{aligned}
c_1 &= \frac{\epsilon\bar{a}}{\epsilon + \bar{f}_{\min}^2},\\
c_\infty &= \beta_1 + \beta_2\Delta + \beta_3\delta,\\
\beta_1 &= \epsilon\sqrt{n}\bar{w} + \bar{f}\bar{h}\bar{w}\sqrt{nN} + \bar{f}\bar{v}\sqrt{n(N+1)},\\
\beta_2 &= \bar{f}\sqrt{nn_y(N+1)}, \quad \beta_3 = 0.5\bar{f}\sqrt{nn_y(N+1)}.
\end{aligned}$$

Proof Taking norms to both sides of the error dynamics (11.17) and exploiting triangular inequalities, we have

$$\begin{aligned}
\|e_{k-N}\| \leq{}& \|(\epsilon I + F_N' F_N)^{-1}\|\\
&\times \Big(\epsilon\|\bar{A}e_{k-N-1}\| + \epsilon\|w_{k-N-1}\| + \|F_N' H_N w_{k-N}^{k-1}\|\\
&+ \|F_N' v_{k-N}^k\| + \|F_N' \tilde{m}_{k-N}^k\| + \|F_N' q_{k-N}^k\|\Big).
\end{aligned}$$

Obviously,

$$\|(\epsilon I + F_N' F_N)^{-1}\| = \frac{1}{\epsilon + \bar{f}_{\min}^2}.$$

Thanks to the fact that $\|F_N' H_N w_{k-N}^{k-1}\| \leq \bar{f}\bar{h}\bar{w}\sqrt{nN}$ and $\|F_N' v_{k-N}^k\| \leq \bar{f}\bar{v}\sqrt{n(N+1)}$, the following result can be derived:

$$\begin{aligned}
\mathbb{E}\{\|e_{k-N}\|\} \leq{}& \frac{1}{\epsilon + \bar{f}_{\min}^2}\Big(\epsilon\bar{a}\mathbb{E}\{\|e_{k-N-1}\|\}\\
&+ \epsilon\sqrt{n}\bar{w} + \bar{f}\bar{h}\bar{w}\sqrt{nN} + \bar{f}\bar{v}\sqrt{n(N+1)}\\
&+ \bar{f}\mathbb{E}\{\|\tilde{m}_{k-N}^k\|\} + \bar{f}\mathbb{E}\{\|q_{k-N}^k\|\}\Big).
\end{aligned} \tag{11.21}$$

In the light of Lemmas 11.2.1 and 11.2.2, we can obtain that

$$\mathbb{E}\{\|\tilde{m}_{i,k}\|^2\} \le n_y\Delta^2 \text{ and } \mathbb{E}\{\|q_{i,k}\|^2\} \le 0.25 n_y\delta^2.$$

Moreover, as the square root function is concave, according to the Jensen's inequality, one has

$$\mathbb{E}\{\|\tilde{m}_{k-N}^k\|\} \le \sqrt{\mathbb{E}\{\|\tilde{m}_{k-N}^k\|^2\}} \le \Delta\sqrt{nn_y(N+1)}. \tag{11.22}$$

Similarly, the following inequality holds:

$$\mathbb{E}\{\|q_{k-N}^k\|\} \le 0.5\,\delta\sqrt{nn_y(N+1)}. \tag{11.23}$$

Substituting (11.22) and (11.23) in (11.21) results in

$$\mathbb{E}\{\|e_{k-N}\|\} \le c_1\mathbb{E}\{\|e_{k-N-1}\|\} + c_2,$$

where $c_2 = c_\infty/(\epsilon + \bar{f}_{\min}^2)$. It is not difficult to prove that $\mathbb{E}\{\|e_{k-N}\|\}$ is bounded above by ξ_{k-N}, the solution to the following dynamical equation

$$\xi_{k-N} = c_1\xi_{k-N-1} + c_2$$

with the initial condition $\xi_0 = \mathbb{E}\{\|e_0\|\}$. By noting that

$$\xi_k = c_1^k\xi_0 + \sum_{t=1}^{k} c_1^{t-1}c_2,$$

we can derive the following inequality:

$$\mathbb{E}\{\|e_{k-N}\|\} \le c_1^{k-N}\mathbb{E}\{\|e_0\|\} + \frac{c_2}{1-c_1}\left(1 - c_1^{k-N}\right). \tag{11.24}$$

Moreover, since $0 < c_1 < 1$, one has

$$\frac{c_2}{1-c_1}\left(1 - c_1^{k-N}\right) \le \frac{c_2}{1-c_1}.$$

Therefore, it can be seen that the error dynamics (11.17) is stochastically uniformly bounded. Moreover, the bound (11.20) can be easily obtained by taking the limit of (11.24). The proof is complete now.

For the case where the system and measurement noises are zero-mean random variables, i.e., $\mathbb{E}\{w_{i,k}\} = 0$ and $\mathbb{E}\{v_{i,k}\} = 0$, we can state the following corollary.

Corollary 11.3.4 *If* $0 < c_1 < 1$, *then the dynamics of the estimation error* (11.17) *is asymptotically stable in the mean sense, i.e.,*

$$\lim_{k\to\infty} \|\mathbb{E}\{e_k\}\| = 0. \tag{11.25}$$

The proof follows directly from the fact that $\mathbb{E}\{\tilde{m}_{i,k}\} = 0$ and $\mathbb{E}\{q_{i,k}\} = 0$, and thus is omitted here.

Remark 11.3 The above corollary reveals that, for the case where the system and measurement noises are zero mean, the moving-horizon estimate will be asymptotically unbiased thanks to our specific efforts to compensate the signal distortions.

The proposed centralized moving-horizon estimation problem explicitly assumes that the measurements of nodes in the network should be available in the central unit for solving the optimization problem. Unfortunately, since it requires huge communication and computational capabilities, the centralized estimation is in general not practical, especially for a large-scale complex network with lots of nodes. To handle such an issue, we consider the decentralized moving-horizon estimation problem in the next section.

11.4 Decentralized Moving-Horizon Estimation

In this section, we will formulate decentralized moving-horizon estimation problems shown in Fig. 11.3. Different from the problem setup in the previous section, we suppose that there does not exist a central unit in the network. Moreover, the information exchange is only permitted within the neighborhood associated with the topology of the network.

The objective of decentralized moving-horizon estimation problems is to estimate the local plant states $x_{i,k-N}, \ldots, x_{i,k}$, denoted by $\hat{x}_{i,k-N|k}, \ldots, \hat{x}_{i,k|k}$, at each stage $k = N, N+1, \ldots$, in a distributed manner based on the neighboring estimates $\hat{x}_{j,k-N|k-1}, \ldots, \hat{x}_{j,k-1|k-1}$ ($j \in \mathcal{N}_i$), the local history measurements $y_{i,k-N}, \ldots, y_{i,k}$, and a prior prediction $\bar{x}_{i,k-N|k-1}$ of the state $x_{i,k-N}$. In this chapter, the neighboring

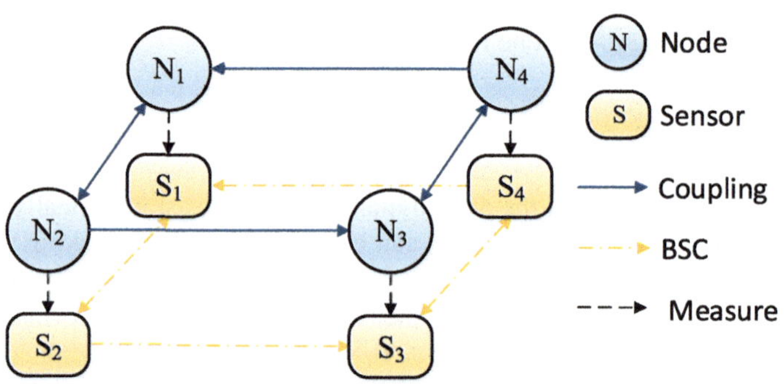

Fig. 11.3 Decentralized moving-horizon estimation

estimates are transmitted under binary encoding schemes. Consequently, the available information of ith node over a finite horizon $[k-N, k]$ is given by

$$I_{i,k-N}^{k} \triangleq \Big(y_{i,k-N}, \ldots, y_{i,k}, m_k^o(\hat{x}_{j,t-N|k-1}, L), \\ \ldots, m_k^o(\hat{x}_{j,k-1|k-1}, L), j \in \mathcal{N}_i \Big).$$

The local predictions $\bar{x}_{i,t|k-1}$ can be determined according to the dynamical equation (11.1) as follows:

$$\bar{x}_{i,t|k-1} = A_i \hat{x}_{i,t-1|k-1} + \sum_{j \in \mathcal{N}_i} b_{ij} \Gamma \tilde{x}_{j,t-1|k-1}, \tag{11.26}$$

for $t = k-N, \ldots, k-1$ and $i = 1, 2, \ldots, n$, where $\tilde{x}_{j,t-1|k-1}$ adopts the following equation to restore $\hat{x}_{j,t-1|k-1}$ just as (11.10).

$$\tilde{x}_{i,t-1|k-1} = \gamma_k^{-1} m_k^o(\hat{x}_{i,t-1|k-1}, L).$$

Accordingly, we can define the equivalent noises as follows:

$$\tilde{m}_{i,t-1|k-1} \triangleq \tilde{x}_{i,t-1|k-1} - m_k(\hat{x}_{i,t-1|k-1}, L).$$

By denoting the truncation error $q_{i,t-1|k-1} \triangleq -\hat{x}_{i,t-1|k-1} + m_k(\hat{x}_{i,t-1|k-1}, L)$, we have

$$\tilde{x}_{i,t-1|k-1} \triangleq \hat{x}_{i,t-1|k-1} + \tilde{m}_{i,t-1|k-1} + q_{i,t-1|k-1}. \tag{11.27}$$

Subsequently, the decentralized moving-horizon estimation problem can be presented as follows.

Problem 2 At each instant k, for given information $\big(\bar{x}_{i,k-N|k-1}, I_{i,k-N}^{k}\big)$, drive the optimal state estimate $\hat{x}_{i,k-N|k}, \ldots, \hat{x}_{i,k|k}$ by minimizing a local least-square cost function defined as follows:

$$J_i(k) = \epsilon \|\hat{x}_{i,k-N|k} - \bar{x}_{i,k-N|k-1}\|^2 \\ + \sum_{t=k-N}^{k} \|y_{i,t} - \hat{y}_{i,t|k}\|^2, \tag{11.28}$$

subject to the following constraints:

$$\begin{cases} \hat{x}_{i,t+1|k} = A_i \hat{x}_{i,t|k} + \sum_{j \in \mathcal{N}_i} b_{ij} \Gamma \tilde{x}_{j,t|k-1} \\ \hat{y}_{i,t|k} = C_i \hat{x}_{i,t|k}, \end{cases} \tag{11.29}$$

for $t = k-N, \ldots, k$, where ϵ is a positive scalar.

Remark 11.4 It should be stressed that the main difference between Problems 1 and 2 lies in that the individual nodes in Problem 2 own computational capabilities, and hence are able to generate local estimates of the target plant by minimizing a local least-square cost function (11.28). In addition, according to the dynamical equation (11.1), we can see that there are complicated couplings between neighboring nodes. Consequently, to calculate $\hat{x}_{i,k-N|k}$, we require the exchange of the information $\hat{x}_{j,k-N|k-1}, \ldots, \hat{x}_{j,k-1|k-1}$ within the neighborhood, which shall be transmitted through BSCs.

Let $v_{i,k-N}^{k} \triangleq \text{vec}\{v_{i,k-N}, \ldots, v_{i,k}\}$ for a generic vector v. Denote $\bar{w}_{k-N}^{k-1} \triangleq \text{vec}\{w_{1,k-N}^{k-1}, \ldots, w_{n,k-N}^{k-1}\}$ and $\bar{v}_{k-N}^{k-1} \triangleq \text{vec}\{v_{1,k-N}^{k-1}, \ldots, v_{n,k-N}^{k-1}\}$. B_i represents the ith row of the matrix B. The rest of the notations are defined as follows:

$$\bar{H}_N \triangleq \begin{bmatrix} I & 0 & \cdots & 0 & 0 & 0 \\ \bar{A} & I & \cdots & 0 & 0 & 0 \\ \vdots & \vdots & \ddots & \vdots & \vdots & 0 \\ \bar{A}^{N-1} & \bar{A}^{N-2} & \cdots & \bar{A} & I & 0 \end{bmatrix}, \quad \bar{F}_N \triangleq \begin{bmatrix} \bar{A} \\ \bar{A}^2 \\ \vdots \\ \bar{A}^N \end{bmatrix},$$

$$F_{i,N} \triangleq \begin{bmatrix} C_i \\ C_i A_i \\ \vdots \\ C_i A_i^N \end{bmatrix}, \quad H_{i,N} \triangleq \begin{bmatrix} 0 & 0 & \cdots & 0 \\ C_i & 0 & \cdots & 0 \\ C_i A_i & C_i & \ddots & 0 \\ \vdots & \vdots & \ddots & \vdots \\ C_i A_i^{N-1} & C_i A_i^{N-2} & \cdots & C_i \end{bmatrix},$$

$$\widetilde{M}_{i,k-N}^{k-1} \triangleq \sum_{j \in \mathcal{N}_i} b_{ij} \, \text{vec}\{\Gamma \tilde{x}_{j,k-N|k-1}, \ldots, \Gamma \tilde{x}_{j,k-1|k-1}\},$$

$$G_i \triangleq I_N \otimes (B_i \otimes \Gamma), \quad \mathcal{G} \triangleq \text{vec}\{G_1, G_2, \ldots, G_n\},$$

$$S_N \triangleq \bar{H}_N (I_N \otimes (B \otimes \Gamma)), \quad \mathcal{A}_N \triangleq \epsilon \bar{A} - \mathcal{F}_N' \mathcal{H}_N \mathcal{G} \bar{F}_N,$$

$$\mathcal{H}_N \triangleq \text{diag}\{H_{1,N}, H_{2,N}, \ldots, H_{n,N}\}, \quad \mathcal{I}_1 \triangleq [0 \;\; I_N \otimes I_{n_x}],$$

$$\mathcal{F}_N \triangleq \text{diag}\{F_{1,N}, F_{2,N}, \ldots, F_{n,N}\}, \quad \mathcal{I}_2 \triangleq [I_{n_x} \;\; 0],$$

$$\mathcal{W}_N \triangleq \mathcal{F}_N' \mathcal{H}_N \mathcal{G}(S_N + \mathcal{I}_1) - \epsilon (B \otimes \Gamma) \mathcal{I}_2.$$

The solution to Problem 2 is presented in the following theorem.

Theorem 11.4.1 *For given information* $(\bar{x}_{i,k-N|k-1}, I_{i,k-N}^{k})$, *Problem 2 has a unique solution given by*

$$\begin{aligned} \hat{x}_{i,k-N|k} = {} & \left(\epsilon I + F_{i,N}' F_{i,N}\right)^{-1} \\ & \times \left(\epsilon \bar{x}_{i,k-N|k-1} + F_{i,N}'(y_{i,k-N}^{k} - H_{i,N} \widetilde{M}_{i,k-N}^{k})\right). \end{aligned} \tag{11.30}$$

Moreover, denote the estimation error of ith node by $e_{i,k-N} \triangleq x_{i,k-N} - \hat{x}_{i,k-N|k}$, *then the augmented error vector* e_{k-N} *satisfies the following dynamical equation:*

$$\begin{aligned} e_{k-N} = &\left(\epsilon I + \mathcal{F}_N'\mathcal{F}_N\right)^{-1}\Big(\mathcal{A}_N e_{k-N-1} + \epsilon w_{k-N-1} \\ &- \mathcal{F}_N'\big(\mathcal{H}_N \bar{w}_{k-N}^{k-1} + \bar{v}_{k-N}^{k}\big) - \mathcal{F}_N'\mathcal{H}_N\mathcal{G}\bar{H}_N w_{k-N-1}^{k-1} \\ &+ \mathcal{W}_N\big(\tilde{m}_{k-N-1|k-1}^{k-1} + q_{k-N-1|k-1}^{k-1}\big)\Big). \end{aligned} \tag{11.31}$$

Proof First, it is not hard to verify that the cost function (11.28) can be rewritten by

$$\begin{aligned} J_i(k) =&\epsilon\|\hat{x}_{i,k-N|k} - \bar{x}_{i,k-N|k-1}\|^2 \\ &+ \|y_{i,k-N}^{k} - F_{i,N}\hat{x}_{i,k-N|k} - H_{i,N}\widetilde{M}_{i,k-N}^{k-1}\|^2. \end{aligned}$$

The necessary condition on the minimum of the above cost function is

$$\begin{aligned} \nabla_{\hat{x}_{i,k-N|k}} J_i(k) &= 2\epsilon(\hat{x}_{i,k-N|k} - \bar{x}_{i,k-N|k-1}) \\ &- 2F_{i,N}'(y_{i,k-N}^{k} - F_{i,N}\hat{x}_{i,k-N|k} - H_{i,N}\widetilde{M}_{i,k-N}^{k-1}) = 0. \end{aligned}$$

Since the Hessian matrix of the cost function is positive definite, the optimal estimate shown in (11.30) is a minimum. In the next step, we will derive the dynamics of estimation error. Utilizing the definition $e_{i,k-N} = x_{i,k-N} - \hat{x}_{i,k-N|k}$, we have

$$\begin{aligned} &(\epsilon I + F_{i,N}'F_{i,N})x_{i,k-N} - F_{i,N}'(y_{i,k-N}^{k} - H_{i,N}\widetilde{M}_{i,k-N}^{k}) \\ &- (\epsilon I + F_{i,N}'F_{i,N})e_{i,k-N} - \epsilon\bar{x}_{i,k-N|k-1} = 0. \end{aligned} \tag{11.32}$$

Denote

$$M_{i,k-N}^{k-1} \triangleq \sum_{\in\mathcal{N}_i} b_{ij}\ \mathrm{vec}\{\Gamma x_{j,k-N}, \ldots, \Gamma x_{j,k-1}\}.$$

Then, it is clear that the measurements of node i from $k-N$ to k can be arranged as follows:

$$y_{i,k-N}^{k} = F_{i,N}x_{i,k-N} + H_{i,N}M_{i,k-N}^{k-1} + H_{i,N}w_{i,k-N}^{k-1} + v_{i,k-N}^{k}. \tag{11.33}$$

In view of (11.27), it can be verified that

$$\begin{aligned} &M_{i,k-N}^{k-1} - \widetilde{M}_{i,k-N}^{k-1} \\ &= G_i\Big(\bar{F}_N e_{k-N-1} - (S_N + \mathcal{I}_1)(\tilde{m}_{k-N-1|k-1}^{k-1} + q_{k-N-1|k-1}^{k-1}) + \bar{H}_N w_{k-N-1}^{k-1}\Big). \end{aligned} \tag{11.34}$$

Substituting (11.26), (11.33) and (11.34) into (11.32), we have the error dynamics as follows:

$$\begin{aligned}
&e_{i,k-N}\\
&= \left(\epsilon I + F'_{i,N}F_{i,N}\right)^{-1}\Big(\epsilon A_i e_{i,k-N-1} + \epsilon w_{i,k-N-1}\\
&- F'_{i,N}\big(H_{i,N}w^{k-1}_{i,k-N} + v^{k}_{i,k-N}\big) - F'_{i,N}H_{i,N}G_i\Big(\bar{F}_N e_{k-N-1}\\
&+ \bar{H}_N w^{k-1}_{k-N-1} - (S_N + \mathcal{I}_1)(\tilde{m}^{k-1}_{k-N-1|k-1} + q^{k-1}_{k-N-1|k-1}\Big)\\
&+ \epsilon \sum_{j\in\mathcal{N}_i} b_{ij}\Gamma(e_{j,k-N-1} - \tilde{m}_{j,k-N-1|k-1} - q_{j,k-N-1|k-1})\Big).
\end{aligned}$$

Augment the above equation from $e_{1,k-N}$ to $e_{n,k-N}$, and then we can get the error dynamics (11.31). The proof is complete now.

Remark 11.5 Theorem 11.4.1 gives a local optimal estimate since every node has a locally known and different cost function (11.28). Moreover, from the results given in (11.30), we can see that all the exploited information, such as the local measurements $y_{i,k}$, neighboring estimates $\tilde{x}_{i,t|k-1}$, and the prediction $\bar{x}_{i,k-N|k-1}$, are available in ith node, and hence the proposed distributed moving-horizon estimator is indeed applicable.

In the sequel, we will investigate the stochastically ultimate boundedness of the estimation error. First, we denote

$$\begin{aligned}
&f \triangleq \|\mathcal{F}_N\|,\ f_{\min} \triangleq \|\mathcal{F}_N\|_{\min},\ h \triangleq \|\mathcal{H}_N\|,\\
&\bar{h} \triangleq \|\bar{H}_N\|,\ \ g \triangleq \|\mathcal{G}\|,\ o \triangleq \|\mathcal{A}_N\|,\ \omega \triangleq \|\mathcal{W}_N\|.
\end{aligned}$$

Based on the aforesaid notations, we present the following theorem.

Theorem 11.4.2 *The dynamics of the estimation error is stochastically ultimately bounded if* $0 < c_1 < 1$. *Moreover, the upper bound of the expectation of the estimation can be given by*

$$\lim_{k\to\infty} \mathbb{E}\big\{\|e_k\|\big\} \le \frac{c_\infty}{\epsilon + f^2_{\min} - o}, \tag{11.35}$$

where

$$\begin{aligned}
&c_1 = \frac{o}{\epsilon + f^2_{\min}},\\
&c_\infty = \beta_1 + \beta_2\Delta + \beta_3\delta,\\
&\beta_1 = \epsilon\bar{w}\sqrt{n} + h\bar{w}f\sqrt{nN} + (\bar{v} + hg\bar{h}\bar{w})f\sqrt{n(N+1)},\\
&\beta_2 = \omega\sqrt{nn_x(N+1)},\ \ \beta_3 = 0.5\omega\sqrt{nn_y(N+1)}.
\end{aligned}$$

Proof Taking norms to both sides of the dynamical equation (11.31) and exploiting triangular inequalities yield that

$$\begin{aligned}
&\|e_{k-N}\| \\
&\leq \frac{1}{\epsilon + f_{\min}^2}\Big(\|\mathcal{A}_N e_{k-N-1}\| + \epsilon\|w_{k-N-1}\| + \|\mathcal{F}'_N \bar{v}^k_{k-N}\| \\
&+ \|\mathcal{F}'_N \mathcal{H}_N \bar{w}^{k-1}_{k-N}\| + \|\mathcal{F}'_N \mathcal{H}_N \mathcal{G} \bar{H}_N w^{k-1}_{k-N-1}\| \\
&+ \|\mathcal{W}_N(\tilde{m}^{k-1}_{k-N-1|k-1} + q^{k-1}_{k-N-1|k-1})\|\Big).
\end{aligned}$$

Notice that $\bar{w}^{k-1}_{k-N}$ and $\bar{v}^k_{k-N}$ can be obtained respectively by reordering the elements of the vectors w^{k-1}_{k-N} and v^k_{k-N}. By choosing an appropriate permutation matrix $\mathcal{P}$, we can show that $\bar{w}^{k-1}_{k-N} = \mathcal{P} w^{k-1}_{k-N}$. According to the properties of permutation matrices, it is obvious that $\|\bar{w}^{k-1}_{k-N}\| = \|w^{k-1}_{k-N}\|$ and $\|\bar{v}^k_{k-N}\| = \|v^{k-1}_{k-N}\|$ which further indicates the following inequalities

$$\|\mathcal{F}'_N \mathcal{H}_N \bar{w}^{k-1}_{k-N}\| \leq f h \bar{w} \sqrt{nN}$$

and

$$\|\mathcal{F}'_N \bar{v}^k_{k-N}\| \leq f \bar{v} \sqrt{n(N+1)}$$

are satisfied. Moreover, similar to the proof in Theorem 11.3.3, one has

$$\mathbb{E}\big\{\|\tilde{m}_{k-N-1|k-1}\|^2\big\} \leq n n_x \Delta^2$$

and

$$\mathbb{E}\big\{\|q_{k-N-1|k-1}\|^2\big\} \leq 0.25 n n_y \delta^2.$$

Therefore, we have

$$\begin{aligned}
&\mathbb{E}\big\{\|e_{k-N}\|\big\} \\
&\leq \frac{1}{\epsilon + f_{\min}^2}\Big(o\mathbb{E}\big\{\|e_{k-N-1}\|\big\} + f h \bar{w} \sqrt{nN} + f \bar{v} \sqrt{n(N+1)} \\
&+ \omega\big(\sqrt{n n_x (N+1)}\Delta + 0.5\sqrt{n n_y (N+1)}\delta\big) \\
&+ f h g \bar{h} \bar{w} \sqrt{n(N+1)} + \epsilon \bar{w} \sqrt{n}\Big).
\end{aligned}$$

It is not difficult to verify that $\mathbb{E}\big\{\|e_{k-N}\|\big\}$ is bounded above by ξ_{k-N}, the solution to the following dynamical equation

$$\xi_{k-N} = c_1 \xi_{k-N-1} + c_2$$

with the initial condition $\xi_0 = \mathbb{E}\{\|e_0\|\}$ where $c_2 = c_\infty/(\epsilon + f_{\min}^2)$. By noting that

$$\xi_k = c_1^k \xi_0 + \sum_{t=1}^{k} c_1^{t-1} c_2,$$

and $0 < c_1 < 1$, we have the following inequality:

$$\mathbb{E}\{\|e_{k-N}\|\} \leq c_1^{k-N} \mathbb{E}\{\|e_0\|\} + \frac{c_\infty}{\epsilon + f_{\min}^2 - o}. \tag{11.36}$$

As such, it can be concluded that the error dynamics (11.31) is stochastically ultimately bounded and the bound (11.35) can be derived by taking limit of (11.36). The proof is complete now.

For the case where the system and measurement noises are zero-mean random variables, similar to Corollary 11.3.4, we can state the following corollary.

Corollary 11.4.3 *If* $0 < c_1 < 1$, *then the dynamics of the estimation error* (11.31) *is asymptotically stable in the mean sense, i.e.,*

$$\lim_{k\to\infty} \|\mathbb{E}\{e_k\}\| = 0. \tag{11.37}$$

Remark 11.6 Theorems 11.3.3–11.4.2 establish the sufficient conditions for the stochastically ultimate boundedness of the estimation error for both the centralized and decentralized moving-horizon estimation strategies. In view of (11.20) and (11.35), we can recognize that the value of such a bound explicitly depends on the length of the bit string as well as the crossover probability.

In the sequel, we will discuss how to choose the length of the bit strings with comprehensive consideration of both energy consumption and filtering performance. A popular model extensively utilized to describe the average transmit power consumption of node i is given by

$$\mathbf{Pow}_i = B_s^2 c_s^2 a_i^2 \left(\ln \frac{2}{p}\right)^2 (2^L - 1)^2$$

where B_s is the transmission symbol rate, $a_i = d_i^\alpha$ is path loss of node i, d_i is the transmission distance, α is the path loss exponent, and $c_s = 2N_f N_0 G_d$. N_f, N_0, and G_d represent the receiver noise figure, the single-sided thermal noise spectral density, and the system constant. Given the targeted filtering performance c_∞^*, we can formulate the optimization problem as follows:

$$\min_{L\in\{1,2,\ldots\}} \sum_{i=1}^{n} \mathbf{Pow}_i^2, \quad \text{s.t. } c_\infty \le c_\infty^*, \tag{11.38}$$

where $c_\infty = \beta_1 + \beta_2\Delta + \beta_3\delta$ with problem-dependent parameters β_1, β_2, and β_3 given in Theorems 11.3.3 or 11.4.2. As the cost function and the filtering performance c_∞ in (11.38) are monotonically increasing and decreasing with respect to length L respectively, the optimal solution can be determined by finding the smallest L such that the inequality $c_\infty \le c_\infty^*$ holds.

11.5 An Illustrative Example

In this section, a numerical example is presented to verify the effectiveness of the proposed centralized/decentralized moving-horizon estimation with binary encoding schemes.

The discrete-time dynamical network (11.1) under consideration consists of $n = 3$ linearly coupled nodes and is with the parameters given as follows:

$$A_1 = \begin{bmatrix} 0.68 & -0.12 \\ 0.21 & -0.66 \end{bmatrix}, \quad A_2 = -\begin{bmatrix} 0.7 & 0.1 \\ 0.2 & 0.76 \end{bmatrix},$$
$$A_3 = -\begin{bmatrix} 0.5 & 0.12 \\ 0.3 & 0.1 \end{bmatrix}.$$

The measurement matrices are given by

$$C_1 = \begin{bmatrix} 0.1 & 0.3 \end{bmatrix}, \; C_2 = \begin{bmatrix} 0.5 & 0.22 \end{bmatrix}, \; C_3 = \begin{bmatrix} 0.5 & 0.11 \end{bmatrix}.$$

The Euclidean norms of the noises $w_{i,k}$ and $v_{i,k}$ are assumed to be bounded by $\bar{w} = 0.02$ and $\bar{v} = 0.1$, respectively. The topology that describes the interconnections of the coupled nodes is shown in Fig. 11.4. The weighted adjacency matrix B is with the entries $b_{12} = b_{23} = b_{31} = b_{32} = 1$, and all the rest entries are zero.

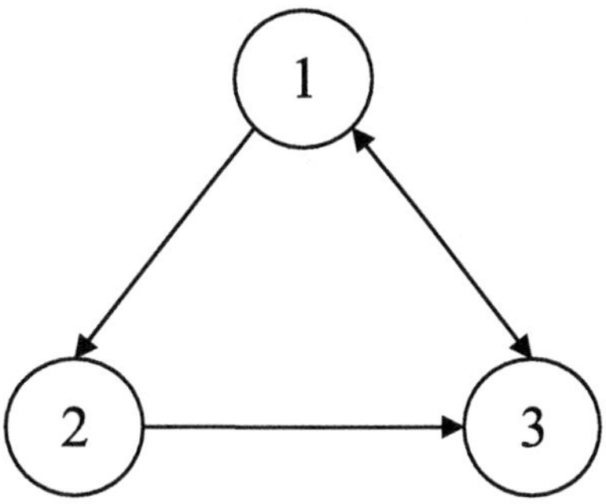

Fig. 11.4 The topology of the coupled network

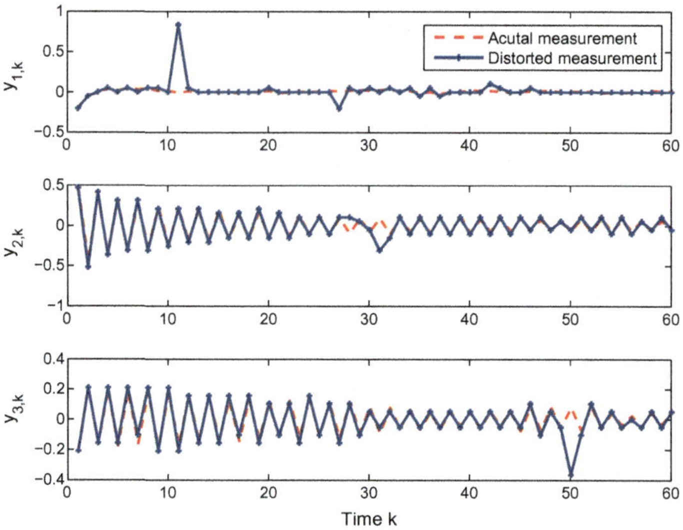

Fig. 11.5 The actual measurements and the distorted measurement via quantized BSCs

Choose the size of the moving-horizon $N = 5$, the range $W = 1$, the length $L = 6$, the scaling weight $\epsilon = 0.2$, and the interval $\delta = 0.01$. The flipping probability p can be identified by statistical experiments in practical applications. It is assumed to be 0.01 here. The initial states of the plants are given by $x_{1,0} = [\,1\ -1\,]'$, $x_{2,0} = [\,0.5\ 1\,]'$ and $x_{3,0} = [\,-0.5\ 0.5\,]'$. Our objective is to derive the optimal estimate for centralized and decentralized moving-horizon estimation, respectively. Suppose that there is no prior knowledge about the initial plant state, so a proper initial estimate could be chosen as $\hat{x}_{1,0} = \hat{x}_{2,0} = \hat{x}_{3,0} = [\,0\ 0\,]^T$.

By resorting to the numerical computing environment Matlab, the optimal centralized moving-horizon estimates are obtained based on (11.16) in Theorem 11.3.1. The simulation results are depicted in Figs. 11.5, 11.6, and 11.7. Figure 11.5 compares the actual measurements $y_{i,k}$ with the received ones $m^o(y_{i,k}, L)$, from which it can be seen that there is significant difference in certain instants due mainly to the bit errors. Figures 11.6 and 11.7 plot the trajectories of the first and second entries of the plant states $x_{i,k}$ and their corresponding estimates. As for the decentralized moving-horizon estimation problem, according to (11.30) in Theorem 11.3.3, the optimal estimates can be derived and the corresponding estimates are shown in Figs. 11.8 and 11.9. All the simulation results have confirmed that the proposed moving-horizon estimators can track the actual states well as expected.

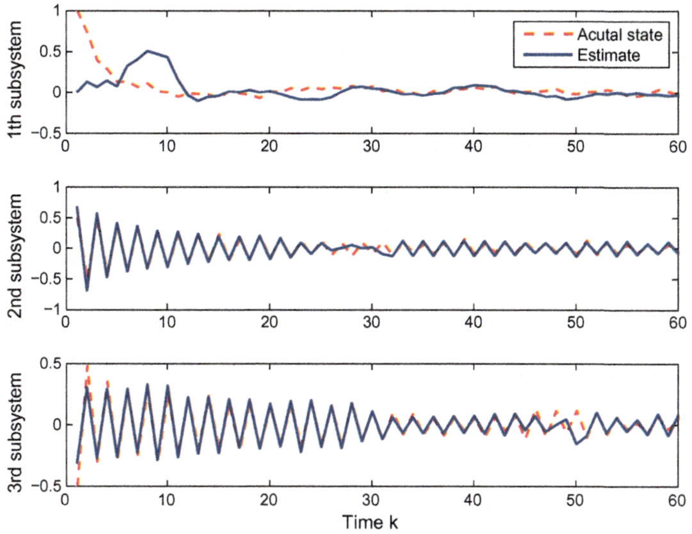

Fig. 11.6 The first entry of the plant states and their estimates based on the centralized moving-horizon estimation

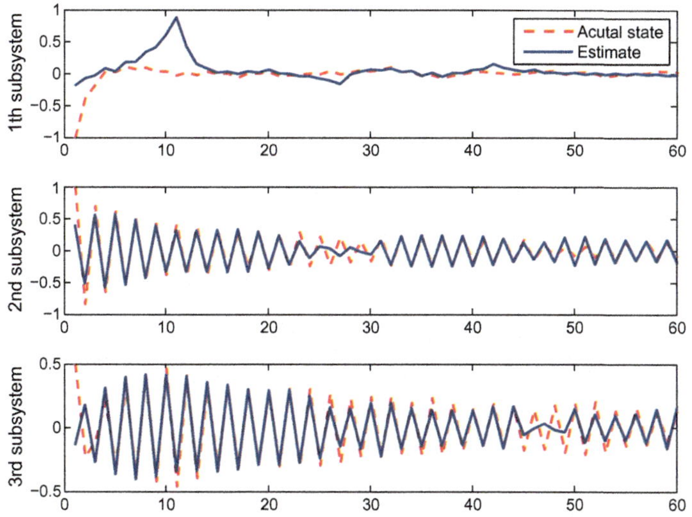

Fig. 11.7 The second entry of the plant states and their estimates based on the centralized moving-horizon estimation

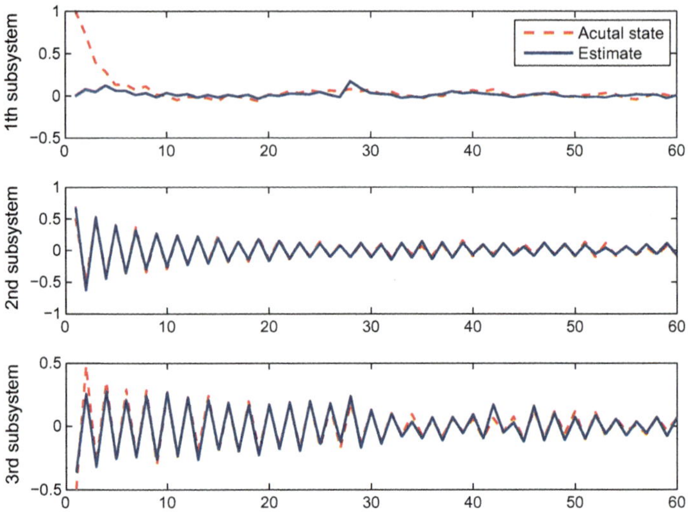

Fig. 11.8 The first entry of the plant states and their estimates based on the decentralized moving-horizon estimation

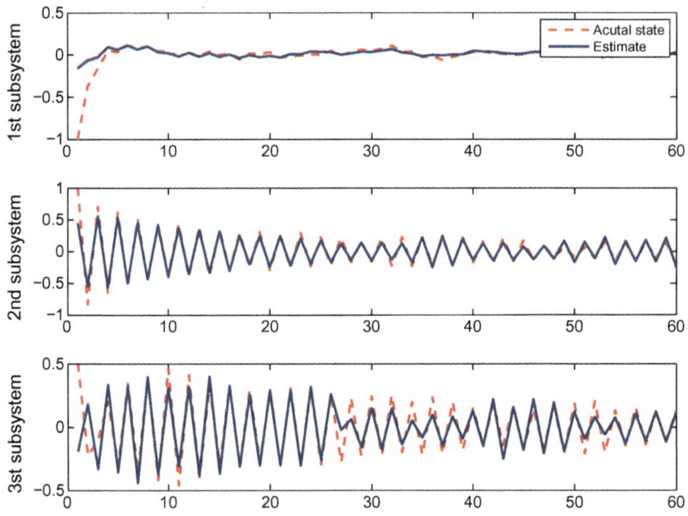

Fig. 11.9 The second entry of the plant states and their estimates based on the decentralized moving-horizon estimation

11.6 Conclusions

In this chapter, the moving-horizon state estimation problems have been investigated for complex networks subject to binary encoding schemes. A novel model has been introduced to quantify the signal errors stemming from the communication processes, and it has been shown that the signal errors can be characterized by a cascade of a multiplicative and an additive white noise associated with the original signal. By solving two specified least-square optimization problems, we have respectively designed the centralized and decentralized moving-horizon estimators. Furthermore, sufficient criteria have been established to ensure stochastically ultimate boundedness of the proposed estimators with the binary encoding scheme. The feasibility and effectiveness of the proposed approaches have been verified via a numerical example.

References

1. H. Su, Z. Rong, M.Z.Q. Chen, X. Wang, G. Chen, H. Wang, Decentralized adaptive pinning control for cluster synchronization of complex dynamical networks. IEEE Trans. Cybern. **43**(1), 394–399 (2013)
2. J. Hu, Z. Wang, S. Liu, H. Gao, A variance-constrained approach to recursive state estimation for time-varying complex networks with missing measurements. Automatica **64**, 155–162 (2016)
3. U.A. Khan, J.M.F. Moura, Distributing the Kalman filter for large-scale systems. IEEE Trans. Signal Process. **56**(10), 4919–4935 (2008)
4. R. Kalman, A new approach to linear filtering and prediction problems. J. Basic Eng. **82**(1), 35–45 (1960)
5. A. Alessandri, M. Baglietto, G. Battistelli, Receding-horizon estimation for discrete-time linear systems. IEEE Trans. Autom. Control **48**(3), 473–478 (2003)
6. C.V. Rao, J.B. Rawlings, J.H. Lee, Constrained linear state estimation-a moving horizon approach. Automatica **37**(10), 1619–1628 (2001)
7. A. Alessandri, M. Awawdeh, Moving-horizon estimation with guaranteed robustness for discrete-time linear systems and measurements subject to outliers. Automatica **67**, 85–93 (2016)
8. C.V. Rao, J.B. Rawlings, D.Q. Mayne, Constrained state estimation for nonlinear discrete-time systems: stability and moving horizon approximations. IEEE Trans. Autom. Control **48**(2), 246–258 (2003)
9. Y. Guo, B. Huang, Moving horizon estimation for switching nonlinear systems. Automatica **49**(11), 3270–3281 (2013)
10. A. Liu, L. Yu, W. Zhang, M. Chen, Moving horizon estimation for networked systems with quantized measurements and packet dropouts. IEEE Trans. Circuits Syst. I Regul. Pap. **60**(7), 1823–1834 (2013)
11. B. Xue, S. Li, Q. Zhu, Moving horizon state estimation for networked control systems with multiple packet dropouts. IEEE Trans. Autom. Control **57**(9), 2360–2366 (2012)
12. H. Leung, C. Seneviratne, M. Xu, A novel statistical model for distributed estimation in wireless sensor networks. IEEE Trans. Signal Process. **63**(12), 3154–3164 (2015)
13. A. Bernstein, K. Steiglitz, J. Hopcroft, Encoding of analog signals for binary symmetric channels. IEEE Trans. Inf. Theory **12**(4), 425–430 (1966)

Chapter 12
Conclusion and Further Work

The focus of this book has been placed on the stochastic control and filtering for networked systems under constrained communication networks. First, the concepts of NCSs have been introduced and the up-to-date research results on analysis and synthesis of networked systems with various constraints have been reviewed. Then, in the following two chapters, the feedback stabilization of networked systems under SNR constraints and the H_∞ control of multi-agents systems under bandwidth constraints have been studied. Subsequently, the networked-based state estimation problems have been taken into account with a particular care for the distributed state estimation. The analysis and synthesis of the distributed filtering problems have been investigated, and sufficient conditions for the existence of distributed filters with desired performance have been obtained in an analytical form.

Generally speaking, this book has established a unified theoretical framework for analysis and synthesis of the networked systems while addressing difficulties induced by networked factors such as packet dropouts, communication bandwidth constraints, and SNR constraints. It should be, however, stressed that the established results are still limited. Some of the related topics for future research are listed below:

- The majority of results about event-based control/filtering problems usually make an explicit assumption that the transmission rules can be established in advance, based on which the controller/filter is designed to ensure certain performance of the systems. It will be interesting to seek a novel method to codesign both the controller/filter and the transmission rules in order to further improve the system performance.
- The information theory has established rigorous definition on the capacity limit of a specific network model which dictates the maximum data rates that can be transferred. Another research frontier is to investigate whether the controller/filter can be implemented to achieve the desired performance with such capacity limits.

Q. Liu et al., *Stochastic Control and Filtering over Constrained Communication Networks*, Studies in Systems, Decision and Control 178,
https://doi.org/10.1007/978-3-030-00157-5_12

- In order to make the data transmission more efficient and reliable, the encoding–decoding techniques are utilized to pretreat and restore the data in the communication process. Note that such techniques inevitably result in the loss of information, and thus considering the impact of encoding–decoding techniques on the networked control/filtering is also an interesting topic.

Printed by Printforce, the Netherlands